The Material Mind

BSPS | OPEN

BSPS Open SERIES

ISSN 2564-3169 (Print) ISSN 2564-3177 (Online)

BSPS Open publishes landmark, cutting edge works that represent the full breadth and diversity of the philosophy of science. Diamond Open Access, all books in this series are available freely to readers everywhere.

Published in collaboration with the British Society for the Philosophy of Science.

No. 1 *The Material Theory of Induction*
John D. Norton

No. 2 *The Large-Scale Structure of Inductive Inference*
John D. Norton

No. 3 *The Material Mind: Reduction and Emergence*
Max Kistler

UNIVERSITY OF CALGARY
Press

The Material Mind
Reduction and Emergence

MAX KISTLER

BSPS OPEN

BSPS Open SERIES
ISSN 2564-3169 (Print) ISSN 2564-3177 (Online)

University of Calgary Press
2500 University Drive NW
Calgary, Alberta
Canada T2N 1N4
press.ucalgary.ca

LIBRARY AND ARCHIVES CANADA CATALOGUING IN PUBLICATION

Title: The material mind : reduction and emergence / Max Kistler.
Other titles: Esprit matériel. English
Names: Kistler, Max, author
Description: Series statement: BSPS open series, 2564-3169 ; no. 3 | Translation of: L'esprit matériel : réduction et émergence. | Includes bibliographical references and index.
Identifiers: Canadiana (print) 20240511301 | Canadiana (ebook) 20240511360 | ISBN 9781773856056 (hardcover) | ISBN 9781773856063 (softcover) | ISBN 9781773856087 (PDF) | ISBN 9781773856094 (EPUB) | ISBN 9781773856070 (Open Access PDF)
Subjects: LCSH: Materialism. | LCSH: Reductionism. | LCSH: Emergence (Philosophy) | LCSH: Mind and body.
Classification: LCC B825 .K5713 2025 | DDC 146/.3—dc23

The University of Calgary Press acknowledges the support of the Government of Alberta through the Alberta Media Fund for our publications. We acknowledge the financial support of the Government of Canada. We acknowledge the financial support of the Canada Council for the Arts for our publishing program.

Alberta Government Canada Canada Council for the Arts Conseil des Arts du Canada

Copyediting by Dallas Harrison
Cover image: Colourbox #48745778
Cover design, page design, and typesetting by Melina Cusano

Contents

List of Figures

Foreword and Acknowledgements

This book is the result of a long project beginning in 2002. A new version of the French edition, first published in 2016, appeared in 2023.

Many people have helped me to clarify my ideas about the place of the mind in nature and about the concepts of reduction, emergence, and disposition on which my analysis is based. Of those who helped me to improve my arguments, I would like to mention François Athané, Alexander Bird, Tim Crane, Javier Cumpa, Alexandre Guay, Paul Noordhof, Robert Kirk, Georges Jobert, Alex Manafu, Thomas Pradeu, and Jürgen Schröder as well as my colleagues in the philosophy departments of Université Paris X-Nanterre, Université Pierre Mendès-France in Grenoble, Université Paris 1 Panthéon-Sorbonne, Institut Jean Nicod, Institut d'histoire et de philosophie des sciences et des techniques, and Centre d'histoire des philosophies modernes de la Sorbonne. I am much indebted to Reinaldo Bernal Velasquez, who read a version of the entire manuscript and provided me with numerous critical comments; to Etienne Ligout, Yanis Pianko, and Vincent Ardourel for their generous advice concerning physics; and to Duncan Lee and Dallas Harrison, who helped me to improve the English version.

I would like to thank the French National Centre for Scientific Research (CNRS — Centre national de la recherche scientifique) for funding my two-year research stay at Institut Jean Nicod from 2002 to 2004; the French National Research Agency (ANR — Agence nationale de la recherche) for funding my research project "Metaphysics of Science" between 2012 and 2015; Durham University (UK) for funding the "Emergent Phenomena in Materials" project (2014–15) as part of the Durham Emergence Project; and Institut Universitaire de France for funding a partial research leave during the preparation of the manuscript.

I am grateful to David Teira for encouraging me to submit the book manuscript to BSPS Open and for his generous support. Warm thanks to Véronique Antomarchi for sharing everything.

Introduction

According to the Western philosophical tradition, we have a dual nature: we are both spiritual and corporeal beings. It feels natural to us to classify our properties into two categories: body and mind. On the one hand, our weight is a bodily attribute; I have weight because I have a body. On the other hand, recalling the sea bath that I took the past summer is a mental attribute; I can recall it because I have a mind. The problem of the relationship between body and mind arises from the dual conception of human nature. Once the duality of attributes has been established, a question inevitably arises: how can body and mind act on each other? Traditional metaphysical doctrines explore different ways of conceiving of mind and body in order to reconcile three seemingly incompatible convictions.

First, both mind and body are real. We have reason to think so because both have causal effects. The weight of my body moves the needle on my weight scale; my memory of the sea bath makes my vocal cords vibrate when I tell a friend about it. This argument for the reality of both mind and body presupposes a traditional metaphysical principle found in Plato, which can be called the "causal criterion of reality": everything that is real is capable of having a causal influence, and everything that has a causal influence is real.

Second, body and mind are radically different because they obey different logics: the body is subject to the laws of nature because of its physical properties. I am attracted to the mass of the Earth like any other massive object. Yet the processes that take place in our minds seem to be exempt from physical laws, obeying rules of a completely different nature. My remembering obeys only a logic of association of ideas; the words to which it gives rise are chosen (most often unconsciously of course) according to criteria of rationality. This allows me to express myself in a language that I believe my friend understands: I choose words that I believe will give her a faithful representation of the events that I have experienced; I omit things that are obvious or that I think my interlocutor already knows in order to focus on information that I think is new to her. The link between my words and the memory that they express therefore obeys criteria of rationality rather than laws of nature.

Third, despite the difference between their natures, body and mind interact. When I recount my memory, my vocal cords vibrate: the mind acts on the body. In the physical world, the breaking of ocean waves on the shore acts on the body, for example by causing the sensation of the sound of these waves, and it is this sensation that plays an essential role in the causal chain that leads to my memory. The physical world and the body, therefore, appear to act on the mind.

Property dualism and reductionist monism are two doctrines that aim to reconcile these three theses and that I will consider in this work. Property dualism holds that mental properties are fundamentally *irreducible* to brain properties. There are two major reasons for thinking that cognition is irreducible to the physical sciences. First is the aforementioned heterogeneity of the criteria for attributing mental properties and physical properties and the corresponding heterogeneity of the norms of correction of these attributions and the explanations in which they are used. Specifically, the attribution of mental properties obeys the norms of practical rationality, whereas physical properties obey the norms imposed by the logic of scientific explanation. Second is the *multi-realizable* nature of psychological properties. Since the 1970s, it has often been taken for granted that the same psychological property can exist in physically different people. If we have taken a sea bath together, then it is possible that we have shared some of our experiences and that we will end up having memories sharing some content, although there might not be any physiological or physical properties that we share that correspond to that shared content.

The problem with property dualism is that it cannot explain the interaction between mind and body. Descartes failed to explain how the thinking substance (i.e., mind) and the extended substance (i.e., matter) can act on each other. Once the radical heterogeneity of the two substances has been established, it is impossible to explain their interaction: if the mind is not in space, then why can my mind act only on my body but not on the bodies of other humans? Why can the mind interact only with the body when the brain is intact? Contemporary property dualism no longer postulates the duality of substances, only the duality of types of property. But this creates an analogous problem: if mental properties and physical properties belong to radically different spheres, then it is hard to see how one can be causally responsible for the other. According to the nomological conception[1] of causation, this requires at least the existence of a psychophysical law of nature, but the second thesis of the radical heterogeneity of the two types of properties forbids the existence of such laws.

The main alternative defended in the current debate on the relationship between mind and body is reductionist materialism, which exists in several variants. Some challenge the radical heterogeneity of physical and mental properties, as expressed in the second thesis above. There are indeed reasons to reject some of the premises of the argument for the irreducibility of psychology to neurophysiology. One reason is that one can doubt the reality of the multi-realization of cognitive properties, at least among the species of animals with which we share some of these cognitive properties. Research within cognitive neuroscience, for example, on the mechanisms underlying vision or memory presupposes that these mechanisms are shared by the various species used in the laboratory for that research. The success of this research justifies the presupposition that many of the neurophysiological mechanisms underlying our visual system are shared by macaques and cats. Furthermore, it justifies that we share certain fundamental mechanisms of memory fixation and learning by conditioning not only with mammals but also with the modest *Aplysia*.[22] Now, if these cognitive properties are not multi-realized, then there is nothing to prevent the existence of psychophysical laws that can be used in a reduction of these cognitive properties.

Some advocates of materialism believe that it is the first thesis above that needs to be sacrificed in order to re-establish the coherence of our conception of ourselves as beings with both a body and a mind. According to eliminativism in its various forms, the second thesis must be interpreted in the sense that mind and body are radically different and even incommensurable conceptual systems. However, the existence of two conceptual systems does not imply the existence of two kinds of properties. At the level of reality, in the sense of causal efficacy, there are only physical properties in the broad sense

1 The word nomological, which derives from the Greek words *nomos* (law) and *logos* (discourse), is the adjective used to designate what relates to the laws of nature.

2 *Aplysia californica* is a marine slug whose very simple nervous system is a favourite object of study for exploring the neurophysiological mechanisms underlying learning.

that includes neurophysiology. Either all real properties are physical (i.e., they are among the properties studied by physics), or they are reducible to these properties. Insofar as psychological concepts cannot be integrated into a scientific theory reducible to neurophysiology, and indirectly to chemistry and physics, they are concepts that might be useful in practice, but they are not reliable guides for judgments about existence. It is certainly convenient to explain my report by referring to my recalling a memory, but this is not a good reason to believe that such events of recalling memories really exist. Insofar as memories are irreducible to entities recognized by neurophysiology, it is more reasonable to locate the causes of my report about last summer exclusively at the level of neuronal activity in my brain. This reasoning leads to denial of the first thesis of the reality of the mind: the mind exists only as a conceptual system; however, there are no real properties that correspond to psychological concepts.

The reasoning developed in this book will lead to a variant of reductionist materialism. I will come to the conclusion that, contrary to the second thesis, the heterogeneity of mind and body is not absolute. But this judgment will be mitigated by a new perspective: rather than reflecting on the relationship between two domains of properties and regularities, each of which appears to be heterogeneous in relation to the other, I propose to situate mind and body within a whole hierarchy of levels of reality. Persons endowed with cognitive properties are corporeal beings composed exclusively of cells; these cells are composed exclusively of organelles and molecules; the organelles are also composed of molecules and the molecules of atoms. In the final analysis, like any other material object, people are composed exclusively of atoms and nothing else. Atoms — and the more fundamental objects discovered by physics of which atoms in turn are composed — are real and act on other real objects according to their physical properties, which therefore are also real. However, there is no reason to deny the existence or reality of the properties of objects composed of atoms. According to the conception developed in this book, new properties *emerge* in complex compound systems.

According to a traditional understanding of emergence, it is incompatible with reduction; an emergent property defies scientific explanation. The emergentists of the early twentieth century considered, for example, that the properties of chemical compounds were emergent in relation to the physical properties of the atoms of which they are composed. It is not controversial that the red colour of rubies is a *systemic* property (i.e., a property possessed only by whole crystals but not by their atomic components); neither aluminum nor oxygen nor chromium (the components of rubies) is red. But the thesis that this red colour is an emergent property of the ruby crystal is stronger than that: in its traditional sense, it means that it is impossible to predict, from knowledge of the atomic components and their properties alone, that the crystal has the property of being red. However, quantum physics has shown that the classification of many chemical properties as emergent was premature: the impossibility of predicting and explaining them in a reductive way (i.e., on the basis of the physical properties of the physical components of molecules) characterizes only the state of science at a given time. Once physics predicted a chemical property, we knew that it was not emergent in an absolute sense but only appeared to be within the framework of a certain theory. More important for my purposes here is that the evolution of science as a whole toward increasing unification authorizes inductive reasoning about all properties that appear to be emergent at a given moment. The conclusion is that emergence is only provisional and relative to a certain theoretical framework. The properties that appear to be emergent today will (probably) be reduced tomorrow. Their reduction is only a question of time and the development of more powerful theories.

This reasoning is entirely justified. However, rather than showing that there are no absolutely emergent properties, it shows only that the concept of emergence used was too strong: imposing the condition of irreducibility leads to the result that absolute emergence does not exist. However, the fact that we manage to construct reductive explanations of certain high-level phenomena is perfectly compatible with the real existence of these phenomena. The link between emergence and reduction that I will develop in this book is as follows. In Chapter 1, I will propose a model according to which the reduction of a property has two logical components. The first is the discovery of a law of composition. This is a non-causal law that can be explained by more fundamental laws. Such an explanation shows that complex objects with a given structure possess certain systemic properties that only appear in structures of this type. The second component is the existence of a structural analogy between the theory deduced from the reducing theory and the theory that preceded the reduction. I will show in Chapter 2 that the discovery of a law of composition is empirical. I will defend this thesis against the influential doctrine of "conceptual reductionism," according to which it is possible to construct the concepts of complex objects a priori, only on the basis of conceptual analysis and knowledge of the microphysical level. In principle, all of the properties of objects belonging to a given level of the hierarchy that reaches from the microscopic to the macroscopic are reducible in this sense. However, the reduction of the properties characterizing a level (e.g., that of macroscopic solids such as crystals) to the level underlying it (e.g., the level of the atoms that make up the crystal) does not remove the reality of the reduced properties. Reduction is not equivalent to identifying the reduced property with the reducing property.

The conception of reduction developed in Chapters 1 and 2 will serve as the basis for a conception of emergence compatible with reduction. This conception will be the subject of Chapter 4. The concept of emergence remains essential to the foundation of the hierarchical conception of reality: it is used to explain why each level possesses new properties (i.e., properties that do not belong to any lower level).

This framework will provide the means to consider the relationship between the mind and the body from a new angle. Cognitive systems, both animal and human, are complex systems with precise structures. Cognitive neuroscience performs the same task of unifying knowledge as physical chemistry or molecular biology: they are sciences "between levels," to use an expression coined by Darden and Maull (1977), built around reductions. The aim of cognitive neuroscience is to explain the systematic appearance of certain cognitive abilities and processes in systems with a certain neurophysiological structure. When it succeeds in explaining the emergence of a cognitive ability, such as the ability to store experiences in long-term memory, it helps to reduce the sense of mystery that the appearance of the mind in nature inspires. In the same way, the process of unifying science as a whole contributes to reducing our lack of understanding of the multiplicity of levels of reality: the reductive explanation in physical chemistry of the chemical bond that causes hydrogen molecules to emerge from hydrogen atoms, or of the colour of a ruby from the electronic structure of this crystal, contributes to reducing our lack of understanding of the existence of a chemical level of properties and laws above the physical level. The reductive explanation of chemical properties and laws based on physics also indirectly helps to reduce the mystery of the emergence of the mind from a body equipped with a sophisticated nervous system. This is because it makes the relationship between the mind and the body appear to be just one of many cases of relationships between properties and laws belonging to adjacent levels in the hierarchy that extends from elementary particles to the mind. The emergence of the mind from the body obeys the same logic of determination according to non-causal laws of composition as the emergence of crystals from their atomic structure.

I have kept the term "emergence" to characterize the relationship between the properties and laws at one level of reality and the levels below it. This might appear to be a contradiction of terms in the context of an analysis that considers that emergent properties and laws are all reducible in principle to the properties and laws of lower levels. However, this terminological choice is justified by the persistence of an element of mystery. The reductive explanation of a property gives us the means to predict and explain its presence, based on the knowledge of the laws of composition and the properties of the underlying level. In this sense, their existence is not truly mysterious. However, there remains an unavoidable element of mystery: insofar as the laws of composition are not laws of reason, we cannot deduce them a priori. These laws are necessarily the subject of empirical discovery. It is true that the progress of science makes laws become progressively incorporated into theories. In such a theory, many "experimental" laws, initially discovered by pure induction from observations, are transformed into theorems. It is possible to understand them insofar as they can be deduced from theoretical axioms. However, by the very definition of the concept of axiom, an axiom cannot be deduced. So, with regard to axioms, there is no answer to the question why does the law have this form rather than another? This is the ultimate and inescapable remnant of mystery that nature retains even when its scientific explanation is complete. The mystery of emergent properties has no other source. It is nevertheless a limit to our ability to understand. This justifies keeping the term "emergence" to characterize the relationship between properties and laws belonging to adjacent levels of reality. Even a complete reductive explanation does not make this relationship absolutely transparent to reason. The residue of incomprehension that we feel when faced with a phenomenon, particularly a cognitive phenomenon, even when its reductive explanation is complete, has its ultimate source in the fact that the fundamental laws of nature cannot be discovered by conceptual analysis a priori but only through experience.

I will arrive at a conception that gives the mind (i.e., our cognitive properties and the laws that they obey) a reality of their own: when I recount a memory to my friend, the mental representations of the sea bath and the desire to share these experiences are among the causes of the vibration of my vocal cords. Now the thesis that mental properties have their own causal efficacy, different from that of the underlying neurophysiological properties, faces a powerful argument developed in particular by Jaegwon Kim (1998). Kim seeks to show that it is impossible for a mental property to influence physical events causally. Insofar as the vibration of my vocal cords is a physical event, its causes can only be physical. The aim of Chapter 5 is to show that this argument is contestable and that there is a way of escaping the verdict that the mind is merely an "epiphenomenon" — something that has no effect at all. Shadows can help to illustrate the concept of an epiphenomenon. When the sun casts a shadow of me while I walk beside a wall, this shadow is an epiphenomenon of my passage because, although it is caused by my presence, it has no causal influence on the subsequent stages of the shadow's appearance on the wall. In this sense, to say that cognitive properties have no causal power is to take them to be epiphenomena. The terms in our mental vocabulary express concepts but not real and causally efficacious properties.

The conception of the mind developed in this book can be plausible only if there is a flaw in Kim's (1998) argument. In fact, Kim's argument for the epiphenomenal character of mental properties, in other words for their being unreal in the causal sense, is simply the application of a more general argument for the unreality of dispositional properties. Indeed, most mental properties — with the possible exception of certain qualities of subjective experience — are dispositional. The fact that I remember my sea bath gives me the disposition to produce a narrative about that bath. The fact that I have learned that there is (normally) someone at the door when the doorbell rings gives me the disposition to behave appropriately when I hear the doorbell ring. One argument against the reality of dispositions is to show that the manifestations of a disposition are always caused by what is called the "causal basis" of the disposition. The causal basis of my disposition to open the door when I hear the doorbell ring is a certain state of the neurons in my brain. The cause of my act of opening the door is not the fact that I know what the sound of the doorbell means but the state of the neurons that act as relays in the causal chain that runs from the fact that I hear the doorbell ring to my act of opening the door. In Chapter 3, I will attach great importance to refuting this general argument against the reality and causal efficacy of dispositional properties. This is justified by its importance for the question of the reality of the mind.

1

Unity of Science and Reduction

1. Introduction

Science aims to broaden and improve our knowledge of the world. Part of this knowledge consists of *descriptions* of things that exist, of events that happen, of processes that take place. But often we are not content with descriptions or with facts. We ask science to help us understand why things are as they are and why events and processes happen. There are two ways of providing us with such an understanding: by discovering the *properties* that objects, events, or processes possess and by knowing the *laws* that they obey by virtue of these properties. These two types of discovery go hand in hand: to hypothesize the existence of a property is to hypothesize the existence of a law (or a set of laws) that imposes constraints on the behaviour or evolution of what has the property. Making the hypothesis of the existence of a law means making progress in our understanding of the world in two ways. The law enables us to complete our knowledge of the properties implied by the law, and it enables us to understand the origin of the links between different things, facts, and events.[1] These links are manifested in regular associations of properties and regular successions of events. The link between properties and the laws in which they appear is more intimate than the empiricist tradition recognizes: possessing a natural property makes it necessary to obey the

1 Causey (1977, 17) counts laws, along with facts, among the objects of knowledge, whereas explanations and theories provide us with understanding. This is not necessarily incompatible with my assertion that laws already enable us to understand why regularities occur. Scientific understanding of the world is an iterative process, and what at a given moment is the object of knowledge can become the starting point for a new interrogation of its why.

laws in which it participates because these laws are constitutive of the identity of the property.[2]

But the discovery of properties and laws is only one step, albeit an essential one, toward a satisfactory understanding of nature. By postulating the existence of new properties that often are not directly observable, and new theoretical laws that these properties obey, theories deepen our understanding. According to the traditional view (shared by Duhem [1906] and logical empiricism), theories provide a unified understanding of a whole field of phenomena by explaining laws first established on the basis of experience and induction. The theory makes it possible to deduce laws previously discovered separately, in particular laws related to observable properties. These laws are called "experimental" (to refer to their origins in the experimental observation of regularities) or "phenomenological" (to point out that the properties concerned by these laws are at least partially observable) to distinguish them from theoretical laws.

A law gives us a unified understanding of a multitude of events. Hooke's Law states that the force exerted by a spring is proportional to its extension.[3] By making these two properties of extension and force appear to be linked by the law, the regularity of their association is understood as arising from a relationship between the properties themselves.[4] The singular events concerning springs can be explained on the basis of the hypothesis that they "fall under" this law or that they are "covered" by it.

The history of science has led not to the creation of a single theory but to a multitude of theories. Sometimes different theories deal with the same objects or phenomena. Gases are the subject of both thermodynamics and classical mechanics. Thermodynamics deals with the regular relationships between the macroscopic properties of gases, such as their temperature and

---

2 This thesis is developed in Kistler (2002a, 2005a). The apparent contingency of the laws is explained by the imperfection of the observed regularities. In terms of manifest properties, regularities are not perfect, such that there are exceptions. Albino crows are not black. The fall of an apple through the air is not uniformly accelerated. The Earth's orbit around the Sun is not a perfect ellipse. Laws impose constraints on properties that are not always manifest separately and that can be described as "capacities" or "dispositional properties." Only the results of their superpositions are manifest. The manifest properties resulting from these superpositions do not always show perfect regularity. I will develop this theme in Chapter 3.

3 This can be expressed concisely in a formula: $F = -kx$, where F represents the force exerted on the spring, k the spring constant, and x the extension of the spring.

4 This conception of laws is developed in Kistler (1999c, 2006d).

pressure, and the volume that they occupy. Classical mechanics, conversely, describes the regularities that govern the behaviour of the molecular components of the same gases. The need for understanding that drives scientific research prevents us from being content with simply juxtaposing the laws discovered by these two theories. After all, there is only one object with one behaviour. How is it that the laws discovered at the component level (in classical mechanics) and at the macroscopic gas level (in thermodynamics) do not contradict each other? The best way to deepen our understanding of this description of objects at several levels, by several theories, is to *reduce* one of the two theories to the other. Such a reduction, the logic of which we will study later, makes it possible to explain how an object can evolve both in accordance with laws that apply to the object as a whole and in accordance with laws that apply to its parts.

The overall aim of my inquiry is to assess the prospects of reducing cognitive psychology to neuroscience. My working hypothesis is that the conceptual problems that arise in this particular case are not fundamentally different from those that arise in the context of other reductions from one theory to another.

2. Deductive and Ontological Unification

One of the aims of science is to deepen our understanding of natural phenomena. When it is discovered that one theory can be reduced to another, this is an important step toward this goal: such a reduction shows that phenomena that have been explained by two independent theories have a common origin. Even if the reduction preserves the existence and some autonomy of the reduced theory, the reduction shows that the phenomena described by the reduced theory are not heterogeneous with respect to the phenomena described by the reducing theory. In this way, reduction gives rise to a "unification" of two previously disconnected domains of knowledge and explanation.

Accepting physicalism gives us a reason in principle to expect the discovery of reductions. According to materialism, everything that exists is material or composed solely of material constituents. *Physicalism* is a contemporary form of materialism, according to which everything that exists is composed exclusively of physical objects. A physical object is one whose properties are all physical, in the sense that they are properties whose identities are determined by the laws of physics. Fields, for example, are physical objects but do not meet the traditional criteria of "material" entities. In this book, I accept physicalism as an empirical hypothesis justified by the success of science as a whole. If physicalism is true, then all theories are about physical objects. This means that

(1) all existing objects have only physical objects as parts and that
(2) all existing objects have properties of only two kinds:
    (a) physical properties and
    (b) properties determined by the physical properties of the object and the properties of its components.

It is useful to distinguish between *explanatory* and *ontological* unification. Each reduction gives rise to an explanatory unification. The Newtonian theory of gravitation, for example, provides a unique framework for explaining both the free fall of a massive body near the Earth's surface and the orbit of a planet around the Sun. Before the Newtonian reduction, these two explanations required the resources of two independent theories — the Galilean theory of free fall and the Keplerian theory of planetary motion — whereas only one was needed afterward. This simplification of the premises necessary to explain apparently heterogeneous phenomena deepens our understanding of phenomena because it makes them appear to belong to a single type (see Kitcher 1989). Reduction consists of showing that all of the constituent laws of the reduced theory derive (i.e., can be deduced) from the laws of the reducing theory, together with certain initial conditions.[5] For example, the reduction of the first Keplerian law of planetary motion (according to which planetary orbits have elliptical shapes) by the Newtonian law of gravitational attraction is based on three presuppositions. First, in the context of calculating their gravitational interactions, the Sun and the planets can be taken to be unextended points, with their masses situated at those points. Second, the force of gravitational attraction between two point masses is a central force that decreases as the inverse of the square of their distance. Third, the only force determining the orbit of planet *m* is the gravitational force between the Sun and *m*.

5 I will return to this derivation later in this chapter.

The fact that reductions always lead to explanatory unification is not controversial. However, it is difficult to explain the exact source of this deeper understanding. According to the tradition of logical empiricism, the derivation of reduced laws from the reducing theory must take the form of a deductive-nomological explanation. This is simply a consequence of the more general doctrine that any scientific explanation must take that form. In this model — thoroughly developed by Nagel (1961) — explanatory unification requires deductive unification. Causey (1977) and Hooker (1981) understand the *explanatory* unification achieved by a reduction in terms of *ontological* unification rather than just deductive unification; reduction often simplifies the ontology. Following Quine, we can consider that theories convey ontological commitment: if a theory is true, then all types of entities over which the axioms and theorems of the theory quantify exist.[6] Belief in the existence of entities of these types is justified to the same extent as belief in the truth of the theory. It makes sense to use the criterion of ontological commitment in a less restrictive form than Quine himself does: the mere fact that a scientific theory successfully introduces a property into the description of its models or into its axioms and theorems provides a reason (fallible of course) to believe in its existence.[7] Because of a reduction, the number of types of entity to whose existence one is committed according to the theories accepted decreases with the number of independent theories. Before the reduction of Mendelian genetics to molecular biology, the former included an ontological commitment to the existence of genes as a primitive type of entity. After the reduction, the ontology is simplified; genes are no longer considered a distinct type of entity. According to the new theory, their causal role is played by biological molecules, first and foremost DNA, and by a number of complex mechanisms that enable them to be replicated, to recombine in sexual reproduction, and to express themselves in the phenotype (see Mossio and Umerez 2014). Before the reduction of the temperature of a gas to the average kinetic energy of the molecules that make it up, gases and their properties, such as pressure and temperature, were fundamental entities in whose existence it

6 "We may be said to countenance such and such an entity if and only if we regard the range of our variables as including such an entity. To *be* is to be a value of a variable" (Quine 1976, 199). See also Quine (1939).

7 It is not necessary, as Quine's original criterion suggests, for the axioms of the theory, or for the description of its models, to quantify over the predicates that express these properties. See Kistler (2012, 2016, 2020).

was reasonable to believe, insofar as the thermodynamics of gases was taken to be true. After the reduction of thermodynamics to classical mechanics, the ontology is simplified by the elimination of the gas as a fundamental type of entity, the fundamental entity now being the set of molecules. We will see that it is sometimes difficult — and often controversial — to judge whether such an elimination of a fundamental type of entity is equivalent to its elimination altogether or whether the new theory, having achieved reductive unification, is still ontologically committed to its existence as a "derived" entity.

A reduction achieves its aim of unifying the representation of the world by bringing together two theories that had distinct domains before reduction. This unification is achieved by providing principles that allow conclusions to be drawn about the objects of the reduced theory, based on premises formulated in the language of the reducing theory. From a premise concerning the average energy of the molecules contained in a sample of gas, the reduction of thermodynamics to classical mechanics allows us to draw a conclusion about the temperature and pressure of this sample. However, the latter concepts apply to macroscopic objects and the former concepts to microscopic objects, one smaller than the other by several orders of magnitude. Similarly, the reduction of elementary learning to neurophysiology makes it possible to draw conclusions about an animal's cognitive state of conditioning from information about the state of the animal's microscopic components. For example, it is possible to conclude from a premise that relates to the change in the conformation of Ca^{2+} channels in certain presynaptic axonal endings of nerve cells in an individual of the species *Aplysia californica* that this individual is in a state of habituation or, conversely, of sensitization in regard to the siphon withdrawal reflex following stimulation of its tail.[8] This might seem to be surprising given that the premise concerns microscopic objects (membrane proteins), whereas the conclusion concerns a disposition to the behaviour of a macroscopic animal. How does the reduction bridge the distance between the domains of such disparate objects of discourse?

8 I will develop this example of the reduction of cognition to neurophysiology later in this chapter, in section 7.

3. The Deductive-Nomological Model of Reduction

Let us begin by examining the form that this question takes within the context of the now classic theory of reduction between theories proposed by Nagel (1961). His analysis of reduction presupposes the framework of the deductive-nomological (D-N) approach to scientific explanation that — since its original proposal by Hempel (1942) and Hempel and Oppenheim (1948) — was supposed to cover not only the explanation of particular facts but also the explanation of laws. According to Nagel, the reduction of a theory, called "secondary," to a more fundamental theory, called "primary," is a scientific explanation in the sense of the D-N model. According to that model, this explanation takes the form of a deduction of the laws of the secondary theory from the laws of the primary theory (Nagel 1961, 338).

Nagel distinguishes "homogeneous" and "heterogeneous" reductions. In a homogeneous reduction, the primary (reducing) and secondary (reduced) theories share the same vocabulary for describing objects in their respective domains. In a heterogeneous reduction, the secondary theory contains primitive descriptive terms that do not belong to the vocabulary, primitive or derived, of the primary theory. The reduction of Galileo's laws of the free fall of objects near the Earth's surface to Newtonian laws of mechanics and gravitation is an example of homogeneous reduction: "Although the two classes of motions are clearly distinct, no concepts are required for describing motions in one area that are not also employed in the other" (Nagel 1961, 339). The reducing theory, like the reduced theory, studies the movements of macroscopic bodies.

Yet the reduction of the macroscopic thermodynamics of gases to classical mechanics is a case of *heterogeneous reduction*. Temperature is a fundamental concept required to describe the objects of the secondary science (thermodynamics), but it is not part of the conceptual repertoire of the primary theory (classical mechanics), which describes the movements of the molecules that make up gases and their interactions. The heterogeneity of the descriptive vocabulary and conceptual apparatus of the primary and secondary theories is at the origin of an "acute sense of mystification" (Nagel 1961, 340) when the reduction of one to the other has been achieved. How is it possible to deduce laws that bear on macroscopic objects and describe links between their macroscopic properties, such as temperature and pressure, from laws that deal with an entirely different domain of objects, smaller by several orders of magnitude? How is it possible to establish a reductive link between these theories when the objects belonging to their respective domains do not share the relevant properties (individual molecules can possess neither temperature nor pressure)? From the point of view of logic, such an explanation by reduction seems to be impossible:

> If the laws of the secondary science contain terms that do not occur in the theoretical assumptions of the primary discipline [i.e., if the reduction is heterogeneous], [then] . . . the logical derivation of the former from the latter is *prima facie* impossible. The claim that the derivation is impossible is based on the familiar logical canon that . . . no term can appear in the conclusion of a formal demonstration unless the term also appears in the premises. (Nagel 1961, 352–53)

The existence of heterogeneous reductions is not controversial, so they must be possible. According to Nagel, the reduction of heterogeneous theories appears to be impossible only insofar as their logical reconstruction omits an essential premise: the logical possibility of such a reduction presupposes the introduction of "assumptions of some kind . . . which postulate suitable relations between whatever is signified by 'A' [a term of the secondary science absent from the vocabulary of the primary science] and traits represented by theoretical terms already present in the primary science" (Nagel 1961, 353–54). Such an assumption is necessary in particular for the term "temperature," absent from the reductive theory, which applies to the molecules of a gas. Nagel calls this condition for the possibility of a reduction the "condition of connectability [sic]" (1961, 354). He puts forward two theses of great importance for my purposes.

First, Nagel departs from earlier work on reduction that stipulated that these *linkages* between concepts of the reduced and reductive theories must take the form of universal statements of biconditional form.[9] A statement of "biconditional" form indicates a necessary and sufficient condition, whereas a statement of "conditional" form indicates only a sufficient condition. "If it's raining, I'll take my umbrella" (conditional form with "if" or "if . . . then

9 Including his own earlier work (Nagel 1951) as well as Woodger (1952) and Kemeny and Oppenheim (1956).

. . .") indicates that the fact that it is raining is *sufficient* for my taking my umbrella. But this statement does not say that rain is a *necessary* condition: it is compatible with the fact that I take my umbrella even when it is sunny. Conversely, the statement in biconditional form (with "if and only if . . .") "I'll take my umbrella if and only if it rains" excludes the possibility that I will take my umbrella when it is sunny. The biconditional form, indicated by the expression "if and only if," indicates that the condition (the expression following the "if") is both sufficient and necessary for me to take my umbrella. In the expression "if and only if," "if" expresses the fact that it is sufficient; "only if" expresses the fact that it is necessary.

With regard to the link between the temperature of a gas and the average kinetic energy of the molecules of which it is composed, statement (B) in biconditional form expresses the thesis that the average kinetic energy is a necessary and sufficient condition for the corresponding temperature:

> (B) (for "biconditional") For any sample of ideal gas x, x has temperature T *if and only if* the molecules making up x have average kinetic energy $E_{kin}(T)$ proportional to T.[10]

However, Nagel (1961, 355n5) explains that the existence of conditions of biconditional form, as in (B), is not necessary for reduction. In order to recover the laws of the secondary theory from the laws of the primary theory, it is sufficient to suppose that there is a *conditional* dependence of the macroscopic property on a microscopic property. We will see that this weakening of the conditions imposed on reductions leads to a deep modification of the concept of reduction.

> (C) (for "conditional") For any sample of ideal gas x, *if* the molecules making up x have an average kinetic energy E_{kin}, *then* x has a temperature $T(E_{kin})$ proportional to E_{kin}.

If the relationship is conditional (and not biconditional), then the fact that the molecules have the mean energy E_{kin} is *sufficient but not necessary*

10 This proportionality is expressed in the following formula: $E_{kin} = \overline{e_{kin}} = \frac{1}{2}m\overline{v^2} = \frac{3}{2}kT$, where m is the mass of a molecule, E_{kin} is the mean kinetic energy of a molecule, k is Boltzmann's constant, and $\overline{v^2}$ is the mean square velocity of the molecules. I will come back to this reduction in detail later.

for the gas to have *T*. This means that it cannot be ruled out that something could have a temperature *T* for a reason other than having molecules with the mean energy E_{kin}.

The quantities of energy and temperature are linked by a numerical equality $E_{kin} = {}^3/_2\, kT$. It might therefore seem to be obvious that the dependence between *T* and E_{kin}, which appear on both sides of the identity symbol =, must be symmetrical, such as the biconditional relation (B), which stipulates that each of *T* and E_{kin} is necessary and sufficient for the other. However, this appearance is ungrounded. The equation $E_{kin} = {}^3/_2\, kT$ is neutral with regard to the question of the nature of the dependence between the properties E_{kin} and *T* themselves. The equation simply expresses the fact that the *numerical values* of these quantities are proportional. It would never occur to anyone to suppose that temperature is identical to kinetic energy, multiplied by *2/(3k)*, for the simple reason that it makes no sense to talk about the multiplication of properties.

An important reason for not requiring biconditional but only conditional links is that it allows the model to be applied to the reduction of *multi-realizable* properties. This is the case with temperature. The state of the microscopic components of an object can vary while its temperature remains the same. Moreover, temperature is a property shared by objects of very different compositions. Gases, and other bodies composed of atoms or molecules, have a temperature by virtue of the kinetic energy of the atoms or molecules of which they are composed. But there are objects that are not composed of molecules yet have a temperature: plasma and radiation. The fact that cognitive properties are multi-realizable (i.e., they can exist in animals of different species thanks to different neurophysiological bases) plays a crucial role in Fodor's (1974) argument for the irreducibility of cognitive properties. This argument is aimed more generally at establishing the irreducibility of the laws of what are known as the "special sciences" (i.e., sciences whose fields of application are more restricted than that of physics, which applies to any spatiotemporal object). However, as we will now see, Fodor's argument depends on the distinction between the biconditional form and the conditional form: multi-realizable properties are irreducible only if we require that a reduction presupposes the discovery of a *biconditional* linking principle between the reducing and reduced properties but not if we admit that in order to reduce it is sufficient to discover a *conditional* relation.

The anti-reductionist argument presupposes that linking principles are biconditional laws: "Bridge laws express symmetrical relations" (Fodor 1974, 129). The crucial point is the thesis that there can be no biconditional law between a multi-realizable property and the different properties realizing it. Let us assume that a given psychological property can be realized, at the neurophysiological level, in principle by an infinite set of structures. In this case, there can be no biconditional law linking the psychological property to neurophysiology since its neurophysiological term would be an open disjunction, with an indeterminate number of terms.

Let us take a closer look at the question of whether the form of the linking statements required for a reduction must necessarily be biconditional or whether a reduction can be achieved with conditional linking statements. The answer to this question depends on the answer to another question: does the reduction require the *derivability* of the laws of the reduced theory, or would a weaker relation of *connectability* be sufficient? The derivability of a law means that it is possible to deduce the law from the reducing theory. Yet, to ensure connectability, it is sufficient for there to be laws that establish a link between properties belonging to the reducing theory and properties belonging to the reduced theory. The existence of conditional bridge laws satisfies what Nagel calls "the condition of connectability" (1961, 354).

> A linking principle of conditional form is a law of nature of the form
>
> All N_1 are P,

where N_1 is a predicate of the reducing theory (e.g., neurophysiological) and P is a predicate of the reduced theory (e.g., psychological).[11] In what follows, the capital letter N represents neurophysiological properties, and the capital letter P represents psychological properties, and N^* and P^* refer to the same properties instantiated at a later time.

11 I will use this non-formal expression. The conditional form appears explicitly in the logical form of the statement: it is a universally quantified conditional: $(x)\ (Nx_1 \rightarrow Px)$, where $\rightarrow$ represents the conditional "if . . . then . . . ," which means "for any object x, if x is N_1, then x is P."

Why are laws according to which it is sufficient to have property N_1 in order also to have property P not sufficient to deduce logically a law of the reduced theory from laws of the reducing theory?[12]

Let us assume that "all N_1 are N_1*" and "all N_2 are N_2* et cetera are neurophysiological laws. In Figure 1.1, these laws are represented by the horizontal arrows at the bottom. Let us suppose that the linking principles also have a conditional form: "All N_1 are P," "all N_2 are P," "all N_1* are P*," and so on. These linking principles correspond to the dotted lines (in vertical or oblique direction) in Figure 1.1, which indicate that a neurophysiological property N is sufficient for a psychological property P.

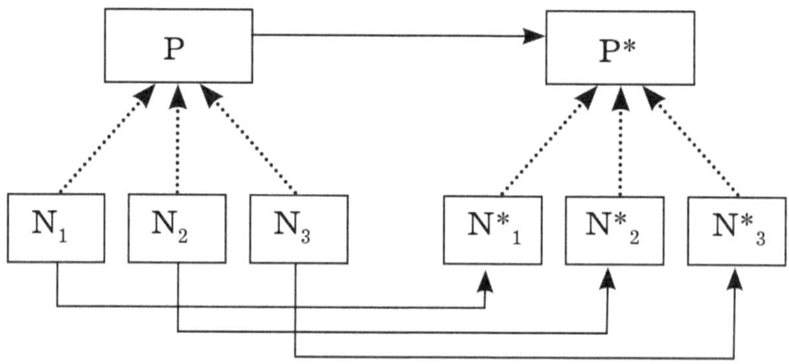

Figure 1.1 Reduction of laws involving multi-realizable properties. Modified from Fodor (1974, 139).

The psychological law "all P are P*" is represented by the top-level horizontal arrow. The question of reduction is to know under which conditions it is possible to deduce logically the psychological law from the neurophysiological laws and the linking principles. It appears that such a deduction is possible only if there are also "top-down" laws corresponding to arrows pointing downward: they would be laws stipulating that it is sufficient to have a certain psychological property P in order to have a certain neurophysiological property N_i. If there is such a "downward" law, for example "all P are N_5," then we can deduce, by transitivity, that

12 For the reasoning that follows, see Marras (2002, 248 ff.).

> All the *P* are N_5 (downward linking law), all N_5 are N_5* (reducing law), all N_5* are *P** (upward linking law), therefore all *P* are *P** (reduced psychological law).

Multiple realizability corresponds precisely to the absence of such downward laws. If the psychological property *P* can be realized, in different individuals, by different neurophysiological properties, N_1, N_2, N_3, et cetera, then having *P* is not sufficient for any one of them in particular.

Richardson is right to note that bridge laws of conditional form are sufficient for connectability, but he is wrong to say that such laws are also sufficient for derivability and therefore for reduction in the strongest sense. As he puts it, "derivability, with its explanatory parsimony, is adequately accounted for . . . if only we find sufficient conditions at a lower level of organization capable of accounting for phenomena initially dealt with at a higher level; and this . . . requires no more than a mapping *from* lower *to* higher level types and *not* a mapping from higher to lower level types" (1979, 548).

Nevertheless, Richardson expresses an important truth: when we discover the microscopic properties and mechanisms, for example biochemical, that underlie a given biological property, we consider that this discovery makes it possible to give a reductive explanation of the phenomenon even if, in other organisms, the microscopic properties and/or the mechanisms are different. Take, for example, the reductive explanation of the "signal" contained in a protein that enables the molecule to reach its destination within the cell. According to the "signal hypothesis," each protein synthesized in the cytoplasm by ribosome and RNA complexes possesses a property that determines its path to its functional destination: this property is the signal that enables the protein to be oriented. It appears that very different microscopic properties can "play the role" of such a signal in different organisms and for different proteins in a given organism. For example, "signal sequences [of amino acids with a protein] for insertion into the ER [endoplasmic reticulum] . . . may vary over 200% in length, apparently show diverse physical chemical interaction with membrane lipids . . . , may or may not be cleaved in serving their function depending on the signal sequence involved, and are sometimes species-specific in their functioning" (Kincaid 1990, 581). The discovery of each of the mechanisms allowing a microproperty underlying the signalling property of a protein to direct the protein toward its destination yields a reductive explanation. In this sense, it is correct to say, with Richardson, that "alternative (possible or actual) mechanisms . . . do not prevent reduction" (1979, 549).

But we must be aware that the reduction achieved by the discovery of microproperties and the associated microscopic mechanisms is not a reduction in the sense of the logical derivability of the higher-level law, in the sense of Nagel. The tension between Richardson's statement that I just quoted and Kincaid's thesis that "such biochemical diversity underlying biological unity is the root obstacle to reduction" (1990, 583) is the result of different interpretations of reduction. Kincaid speaks of reduction in the sense of logical derivation, whereas Richardson's thesis can be defended if the word *reduction* is given a weaker meaning.[13] Insofar as it is sufficiently enlightening to derive, for one or another of the types of system possessing the reduced property, a law structurally equivalent to the reduced law $Sx \rightarrow S^*x$, of the form $P_ix \rightarrow P_i^*x$, from the reducing theory that correctly describes that type of system possessing S, we can consider that such a discovery constitutes a "reductive explanation" though not in the Nagelian sense of derivability.

In this sense, Ausonio Marras observes that it is legitimate to say that biological properties realized in different ways in different types of organisms can nevertheless be reduced, provided that we weaken the Nagelian conditions for reducibility, so that "we take the essential core of the reduction to be *not* the derivation of the *actual laws* of the target theory from the laws of the base theory, but merely the derivation of the *images* of such laws under appropriate boundary conditions" (Marras 2002, 249).[14]

In such a weakened conception of reduction, the fact that an indeterminate number of physical properties N_i underlie a mental property P does not prevent the reduction of P. Multiple realizability is compatible with the possibility of reducing the multi-realizable property to different realizing properties. This change of perspective has several important consequences. First, as Kim has pointed out, such reductions are only "local": "If each of the psychological kinds posited in a psychological theory has a physical realization for a fixed species, the theory can be 'locally reduced' to the physical theory of that species" (1992a, 19; 1993b, 328). Second, the concept of local reduction

13 Richardson (1979) himself does not seem to be aware of this: he wrongly claims that conditional bridge laws are sufficient for Nagelian derivability.

14 In this context, law A is called the "image" of law B if A and B are different but analogous in the sense that A shares (part of) the structure of B.

forces us to abandon the thesis (defended by Causey 1977) according to which the discovery of a reduction necessarily takes the form of the discovery of *the identity* of properties.[15] Conversely, the fact that there is no law of biconditional form on which the reduction is based constitutes a reason to deny that there is a (unique) reducing property identical to the reduced property.

Awareness of the possibility of local reductions is certainly an important step toward anchoring the mind in matter by showing that the multi-realizability of psychological properties does not present an insurmountable obstacle. There is a weaker model of reduction that does not require the deduction of the reduced theory from the reducing theory but only the discovery of local theories that apply to one or more realizations. The fact that the local reducing law $P_i x \rightarrow P_i{}^*x$ can be seen as reducing the law $Sx \rightarrow S^*x$ makes local reduction compatible with an important feature of reductions as they occur in the history of science. Most reductions are accompanied by *corrections* to the reduced theory. According to the model developed by Schaffner and others, what is deduced from the reducing theory is not the old theory that is the subject of the reduction but a new theory that resembles it.[16]

In Schaffner's (1967) terms, the theory $T_R{}^*$ derived from T_B (which stands for "base theory"; it is the reducing theory) must be in a relationship of "close similarity" to the original theory T_R that was to be reduced; the numerical predictions made from $T_R{}^*$ must be "very close" to those made from T_R; moreover, between the theory T_R to be reduced and the theory $T_R{}^*$ actually derivable from T_B, there must be a "strong analogy" or "positive analogy" (Schaffner 1967, 144). The abandonment of the deducibility requirement and its replacement by the requirement of the deducibility of a theory *analogous* to the reduced theory make Schaffner's conception compatible with multi-realizability. In the case of a multi-realizable property, there are several reducing theories, each serving as a basis for the deduction of a theory analogous to the reduced theory, without the different theories thus obtained being identical to each other.

---

15 According to Esfeld and Sachse (2011), the functional reduction of higher-level properties does justice to the existence of special sciences even though these higher-level properties are *locally* identical to physical structures.

16 The model developed by Schaffner (1967, 1993) to account for reductions that do not obey the Nagelian requirement of derivability was taken up by Hooker (1981), Churchland (1985), and Bickle (1998).

4. The Model of Reduction by Analogy

Historically, the new reduction model has been constructed to take account of the fact, noted by many authors, that reductions do not preserve the details of the theories that existed prior to the reductions. On the contrary, an important motivation for the search for a reduction is the corrective modification that the new theory T_B imposes on the old theory T_R, in terms of both observable predictions and theoretical assertions. This observation can lead to two conclusions. According to Popper (1957), Feyerabend (1962), and Kuhn (1962), the fact that "falsifications" or "paradigm shifts" lead to the adoption of a new theory incompatible with the old theory shows that it is not appropriate to speak of a reduction. Rather, it is the "replacement" or "elimination" of the old theory in favour of a radically different new theory. If T_R and T_B are incompatible or even incommensurable, one cannot be reduced to the other insofar as a reduction consists of justifying the old theory on new grounds. Among the cases traditionally referred to as "reductions," it is exceptional for the reduction to lead to the retention of the reduced theory in its precise form.

Popper (1972, 198–200) and Feyerabend (1962, 46–48) show this with the example of the corrective reduction of the Galilean law of free fall.[17] According to Galileo, a projectile launched from the surface of the Earth moves along a parabola. If its initial velocity is zero, then its free fall is a rectilinear and uniformly accelerated motion toward the centre of the Earth. However, this Galilean law cannot be derived as it stands from the Newtonian laws of motion. Newton showed that the trajectory of a projectile is elliptical (in the case of a spherically symmetrical attractor) and never strictly parabolic. However, the Newtonian theory also helps to explain the success of the Galilean law despite its falsity: when the total length of the projectile's trajectory is small compared with the Earth's radius, the parabolic shape is a good approximation to the elliptical trajectory.

The same conclusion can be drawn for Kepler's laws of motion of the planets around the Sun. Kepler's third law states that the ratio of the cube

17 Glymour (1970, 345) and Sklar (1993, 335) point out that, in order to derive Galileo's law, it is necessary to make the counterfactual hypotheses that there are no forces acting on the falling body other than gravitation (no friction in particular) and that the Earth is perfectly spherical. Popper (1972, 200) notes that Galileo's law can be deduced within the framework of Newtonian theory only if a false premise is added: the radius of the Earth is infinite. He adds that this premise is not only de facto false but also without sense since it has absurd consequences in Newtonian theory.

of the planet's mean distance from the Sun, *a*, to the square of the planet's period (i.e., the duration of one revolution around the Sun), *T*, is a constant (i.e., the same for all planets).

(K) $a^3 / T^2 = k$ (where k is a constant)

In Newtonian theory, we can only derive the following law, according to which this ratio, for a system composed of two point masses, is proportional to the sum of their masses m_1 and m_2.

(N) $a^3 / T^2 = k(m_1 + m_2)$ (where k is a constant)

From a Newtonian perspective, Kepler's original law is false for two reasons. First, the law (N) only applies to a system of two bodies and not when there are several planets that also influence each other. Not only is it false that there is only one planet, but also, if there were, then Kepler's law would be meaningless: its content is a regularity in the behaviour of all planets. If there were only one planet, then there would be no point in trying to establish any regularity between the planets. Second, (K) would be true (i.e., $m_1 + m_2$ would be a constant for all the planets) only if the masses of all the planets were the same or negligible compared with the mass, m_1, of the Sun.

According to Schaffner (1967), Churchland (1979, 1985), Hooker (1981), and Bickle (1998), in typical cases of reduction, such as that of the law of free fall or Kepler's third law, T_R is not derivable from T_B in any formal sense, and the primitive terms of T_R have no equivalents (nomologically co-extensional terms) in the language of T_B. In the situation that results from a "corrective" or "approximate" reduction, the new theory T_B can typically explain why the old theory was able to fulfill its explanatory and predictive role, although it is now considered to be false. Placing T_R and T_B in parallel allows us to understand in what sense T_R is an "approximation" of T_B. In some cases, one can indicate fictitious situations in which T_R can be deduced from T_B. One can then "obtain T_R from T_B, deductively: if one conjoins to T_B, certain contrary to fact premises . . . , one can obtain T_R" (Schaffner 1967, 138; variables modified).

In general, reduction leads to a modification of the reduced theory. For example, the reduction of the psychological theory of learning by the neurobiological theory of Kandel (which I will present later in this chapter) has led to a change in the conception of the different types of learning: "Available evidence suggests that classical conditioning and sensitization are not fundamentally different, as is frequently thought, but rather the cellular mechanism of conditioning appears to be an elaboration of the mechanism of sensitization" (Hawkins and Kandel 1984, 389). In other words, "neurobiology may have discovered that simple and associative learning are not as different as psychology has supposed" (Gold and Stoljar 1999, 864). In a similar way, molecular biology's reductive explanation of the biological concept of the gene has led to its modification without, however, eliminating it. The old concept of the gene has been split into three different concepts corresponding to different criteria of gene identity.[18]

Kemeny and Oppenheim (1956) had already observed that the fact that the reducing theory generally corrects the reduced theory, and the fact that the laws derived from the reducing theory are therefore generally incompatible with the laws of the reduced theory, make it impossible to satisfy the requirement imposed on reduction by Nagel (1951) and Woodger (1952) that there be biconditionals linking the vocabularies of the reduced and reducing theories.[19] According to Kemeny and Oppenheim, "any actual example has to be stretched considerably if it is to exemplify connections by means of biconditionals, and most examples will under no circumstances fall under this pattern" (1956, 13). Their article anticipates the central thesis of Schaffner's theory: "We might suggest that it is some modification T_R* of T_R that is actually reduced to T_B" (17; symbols modified). However, they refrain from

18 Genes can be thought of as units of recombination: in this sense, a gene is a "recon." They can also be seen as units of mutation: in this sense, a gene is a "muton." But what corresponds most closely to the traditional functional concept of a gene is what enables hereditary traits to be transmitted from one generation to the next: in this sense, a gene is a "cistron" (Kitcher 1982; Rosenberg 1985). In the words of Endicott, "'cistron' is a corrected image of the Mendelian gene (a term in T_R*, and hence a term supposedly [according to the CHB model, where CHB stands for Churchland, Hooker, Bickle] formulated within the idiom of T_B). Yet it was not created from molecular genetics (T_B) *ex nihilo*, but from the pressure of the original Mendelian theory (T_R) to find a structure with the function of a gene. So *co*-evolved terms within T_B or rather its subset T_R* are by their very nature dually constrained by the rationales and conceptual resources grounded at both levels. In a word, they are theoretical hybrids." (1998, 65)

19 Wimsatt points out that, insofar as one can reconstruct the reduction of an old theory — now considered false — from a new theory that corrects it, as a deductive argument whose form is valid, "there had better be an equivocation somewhere!" (1976a, 218).

developing this idea because they consider that "such a T_R^* is not usually formed, and it may be very difficult to formulate it" (17).[20]

Instead of judging, as Kemeny and Oppenheim do, that correctives are too complex to be subject to formal analysis, and instead of judging, like Popper, Feyerabend, and Kuhn do, that they are not really reductions since they refute the old theory T_R (reduction requires that it be justified by being deducible from a more fundamental theory T_B), several authors — including Putnam (1965), Hempel (1965a), and Schaffner (1967) — have tried to construct a more sophisticated concept of reduction, which makes it possible to account for the fact that the reductions found in the history of science do not preserve the reduced theory but generally correct it.

Schaffner proposes a "general reduction paradigm" (1967, 144) supposed to apply both to reformative reductions that correct the reduced theory and to conservative reductions that preserve it. According to this model, T_B reduces T_R if it is possible to derive from T_B a "corrected" theory T_R^* that "bears a close similarity" to T_R and that produces quantitative predictions that are "very close" to those of T_R (144). Schaffner rejects Nagel's (1961) thesis that binding principles of the conditional — rather than biconditional — form are sufficient to accomplish a reduction. His justification for taking up the earlier criterion formulated by Nagel (1951) and Woodger (1952) is that, following Feigl (1958), the most plausible interpretation of linking statements is that they express "synthetic identities" (145) and that an identity statement is a fortiori of biconditional form. He distinguishes between the association and subsequent identification of objects in the domain of theories T_B and T_R

20 The model of reduction that they propose in exchange does not take account of a direct relationship between the *theories*: for Kemeny and Oppenheim, it is impossible to find a link between the reduced and reducing theories that accounts for the reduction. The only condition that it is possible to express concerns the *observable consequences* of these theories. In stronger reduction models, Kemeny and Oppenheim's condition will be considered necessary but not sufficient. A reduction that satisfies only the minimal condition of Kemeny and Oppenheim does not establish any link between the theories themselves; it is therefore inappropriate to speak of an intertheoretical reduction. In a case in which there is no intertheoretical reduction but replacement (or "elimination"), the requirement of Kemeny and Oppenheim gives precise meaning to the idea that the new theory explains the entire domain of phenomena that the old theory explained. In Schaffner's terms, it is a matter of "reduction as explanation of the experimental domain of the replaced theory. Though in this latter case we do not have intertheoretic reduction, we do maintain the 'branch' reduction" (1992, 320) of some science conceived in terms of the domain of phenomena it explains. Also see Schaffner (1993, 423, 431).

and the association and subsequent identification of the properties expressed by their predicates, stating that "it is possible to set up a one-to-one correspondence representing synthetic identity between individuals or groups of individuals of T_B and T_R*" (144; symbols modified).[21]

This is a major change in the conception of reduction. For Nagel, the reduced and reducing theories possess and retain their own domain of individuals and properties, with the linking principles allowing the deductive integration of the laws of T_R into the theoretical framework of T_B. The linking principles merely express the existence of dependencies that form the basis of the reductive inferences that move from one domain to the other. There is a big step to be taken between the hypothesis that the temperature of a macroscopic gas *depends on* the kinetic energy of the microscopic molecules that make it up and the hypothesis of *the identity* of these two domains, even if it is "synthetic" (i.e., known a posteriori). Then it is just one step further to accept the idea that the deduction that corresponds to a Nagelian reduction, between T_B and T_R*, is in fact an *intratheoretical* deduction that belongs entirely to the reducing theory.[22] It is then the link of analogy between T_R* and the old theory T_R that corresponds to the reduction relation.

21 Schaffner adds that the reduction functions that associate individual constants and predicates are "in general . . . interpretable as expressing referential identity" (1967, 144). See also Schaffner (1976, 618). The transition from function to identity is much less straightforward than Schaffner presents it to be. The existence of an association function between the properties described by T_R and the properties described by T_B is compatible with the thesis of the emergence of the former, according to interaction laws. Schaffner sets as a condition for reduction that "T_R* (entities) = function [T_B (entities)]" and that "T_R* (predicates) = function [T_B (predicates)]" (1976, 618). In the case of a multi-realizable property, there is such a regular association function. But this condition is much more general (or weaker) than the "referential identity" condition that Schaffner seems to consider as equivalent. It corresponds to the case in which the function is identity.

22 This step has been taken by Churchland, Hooker, and Bickle. See, for example, Churchland (1985, 11) and Bickle (1998, 108).

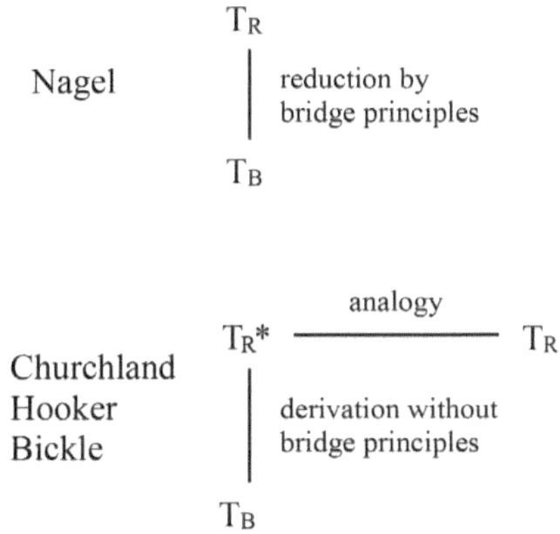

Figure 1.2 Two reduction models, differing with respect to historical change. In Nagel's model, the old theory T_R is deduced from the new theory T_B using linking principles. In the CHB (Churchland-Hooker-Bickle) model, this deduction is carried out without any use of linking principles, using only the conceptual resources of T_B, and leads to a new theory T_R^*, analogous to the reduced theory T_R. T_R^* is of the same theoretical level as T_R but improves it.

This conception aims to assimilate microreduction, the subject of Nagel's model, to the reduction between successive theories dealing with objects of the same size.[23] In the sciences, the term "reduction" is often used to characterize the relationship between a new theory, such as special relativity, and an older theory that it replaces while dealing with the same objects, in this case classical Newtonian mechanics. In this sense, the term "reduction" refers to the relationship between a new theory, T_R^*, and an older theory, T_R, which strictly speaking is false and which T_R^* replaces. In general, T_R can be recovered from T_R^* by giving certain parameters counterfactual values (e.g., an infinite speed of light in the equations of relativistic mechanics). In this context, it is said that "T_R^* reduces to T_R." In this sense, we can say that relativistic mechanics "reduces to" classical Newtonian mechanics "in the limit of small speeds" (i.e., in the limit in which the speeds under consideration are much smaller than the speed *c* of light). For example, in relativistic mechanics, the

23 On this second concept of reduction, see Glymour (1970), Batterman (1995, 2002), and Rueger (2000b, 2001, 2004); on the comparison between the two concepts, see Nickles (1973) and Wimsatt (1976a, 215 ff.).

momentum p of an object is equal to $p = \frac{mv}{\sqrt{1-(v^2/c^2)}}$, where m indicates the mass of the object (defined in a frame in which it is at rest), v its velocity, and c the speed of light. It is commonly said in science that this equation "reduces" to the classical momentum equation, $p = mv$, "in the limit of small speeds": that is, in situations in which $v \ll c$ (the object's speed v is much smaller than the speed of light c).[24] In this type of case, the reduction links two theories, T_R and T_R^*, which have the same field of application. The theory T_R^* replaces the old T_R because it allows its errors to be corrected while reproducing its successes: T_R^* is empirically stronger or simpler, or both, than T_R. But the two theories deal with the same objects: unlike in microreduction, in which the reducing theory is concerned with parts of the objects that are the subject of the reduced theory, this is an "intralevel" or "domain-preserving" reduction (Nickles 1973, 181).

Schaffner's model of reduction (and its variants developed by Churchland, Hooker, and Bickle) aims to assimilate "intralevel" reduction to microreduction or "domain-combining" reduction (Nickles 1973, 181), where the domains of the reduced and reducing theories are at different levels of the micro-macro scale.[25] This seems to be difficult to conceive since the first reduction model requires that the objects of both theories belong to the same domain (and the same level of the micro-macro scale), whereas the second model is supposed to cover reductions between theories describing objects of different sizes, where the domain of the reducing theory relates to objects that are *parts* of the objects in the domain of the reduced theory. How could "interlevel" microreduction be assimilated into "intralevel" reduction?

24 With $\gamma = 1/\sqrt{1-(v/c)^2}$, γm is the relativistic mass (i.e., the mass in a reference frame where the object is at speed v). If $v \ll c$, then $\gamma \approx 1$, and $m(v) \approx m$, so that $p \approx mv$.

25 Rueger (2004) attempts to achieve this assimilation by another means. He sees the microdescriptions and macrodescriptions of a system as two descriptions of the same level (i.e., the macrolevel). The microdescription involves the attribution of a microstructural property in terms of a variable on the microscopic scale. Both are causally efficacious, the macroproperty being a "part" of the "micro"property (which in fact is a macroproperty that takes the microstructure into account). The least that we can say is that the thesis that one property can be "part" of another, in the mereological sense, needs to be justified. In the absence of such a justification, it is the *description* of the macroproperty that is "part" (in the sense of the terms of a conjunction) of the description of the microproperty, where the latter is written in the form of a Taylor series development of the solution of the equation that determines the microproperty.

Schaffner's idea — later taken up by Churchland (1979, 1985), Hooker (1981), and Bickle (1998) — is that the reduction of T_R to T_B comprises two distinct stages. The first stage involves the construction, within T_B and under certain conditions C, of a theory T_R^* that replaces the old theory to be reduced T_R. This first step is supposed to bridge the distance between the microscopic theory T_B and a theory of macroscopic objects, *without* resorting to Nagelian linking or "bridge" principles. It therefore respects the stricter conditions imposed by an a priori reduction model: the higher-level theory is deduced solely from the conceptual resources of the reducing theory, without recourse to linking principles or other conceptual resources foreign to the microscopic theory T_B. I propose to call such a conception of reduction "conceptual reduction." It is only at the second stage that the CHB model of reduction (as noted above, I will use this acronym to refer to the model elaborated by Churchland, Hooker, and Bickle) makes use of "linking principles" between terms of the old theory and terms of the new theory. Between T_R and T_R^*, Schaffner says, there must be a "positive analogy" (1967, 144). In his development of Schaffner's conception, Hooker argues that the existence of an "analog relation" between the theory T_R^* derived from the base theory T_B and the theory to be reduced, T_R, "warrants claiming (some kind of) reduction relation, R, between T_R and T_B" (1981, 49).[26]

Let us assume the existence of two stages, the first of which consists of crossing the distance between the microscopic theory T_B and the macroscopic theory T_R^* by means of a deduction that exploits only the resources of T_B. Insofar as T_R^* is conceived as a theory of the same level as T_R that corrects T_R, the relationship between T_R^* and T_R resembles an intralevel reduction in the sense of Nickles (1973) and Wimsatt (1976a). The controversial thesis required to justify this assimilation concerns the first stage of the reduction. The CHB model sees the derivation of T_R^* from T_B as a deduction internal to T_B. According to this model, it is not necessary to use Nagelian "linking principles" to bridge the distance between the microdomain of T_B and the macrodomain of T_R^*. In this sense, the derivation of T_R^* from T_B is therefore

26 In Churchland's words, "a reduction consists in the deduction, within T_B, not of T_R itself, but rather of a roughly equipotent *image* of T_R, an image still expressed in the vocabulary proper to T_B" (1985, 10; symbols modified). Bickle (1998) gives a detailed account of the conception of reduction developed by Schaffner, Hooker, and Churchland.

intratheoretical and interlevel.[27] To take account of the fact that the reduced theory T_R is often only analogous to a particular case of application of T_B, the derivation of $T_R{}^*$ uses "limiting assumptions" and "boundary conditions" as premises in addition to T_B. These are the types of assumptions regularly used in intralevel reductions. To give an example of a limiting assumption, we can think of the fact that, to find equations structurally analogous ($T_R{}^*$) to the equations of classical mechanics (T_R) from relativistic mechanics (T_B), we have to use the assumption that the speeds of the objects to which the equations are supposed to apply are very slow compared with the speed of light. To give an example of boundary conditions, in the derivation of certain equations of statistical mechanics ($T_R{}^*$) analogous to the thermodynamic equations (T_R) for a gas, it is assumed that the number of molecules in the gas remains constant and that the gas remains confined to a constant volume.

Before going any further, I will examine in the next section whether the model of conceptual reduction suggested by Schaffner, Churchland, Hooker, and Bickle applies to a paradigmatic case of reduction.

5. The Reduction of Thermodynamics to Classical Mechanics

The controversial condition of the CHB model requires that it be possible to deduce, from T_B alone, the laws of $T_R{}^*$ without employing linking laws that appeal to T_R concepts. In order to evaluate the thesis that this model adequately represents historical cases of scientific reductions, and to avoid begging the question, I will examine whether the CHB model can account for a paradigmatic case of a successful reduction, that of thermodynamics to classical mechanics. Bickle (1998) argues that the CHB model passes this test: he tries to show that it is able, in particular, to account for the reduction of the ideal gas law to classical mechanics. According to Bickle, it is possible to derive, within classical mechanics, the following "analog structure" of the ideal gas law:

27 The crucial thesis is that no "linking principle" or "bridge law" is needed to derive the "image" $T_R{}^*$ from T_B (the reducing theory), an image whose isomorphism with T_R (the reduced theory) justifies the claim that T_B reduces T_R. As Churchland puts it, "the correspondence rules play no part whatever in the deduction. They show up only later, and not necessarily as material-mode statements, but as mere ordered pairs: <Ax, Jx>, <Bx, Kx>, . . . " (1985, 10).

$$(*)\quad \frac{Nm\overline{v^2}}{3lwh}\cdot lwh = \frac{2N}{3}\cdot\frac{1}{2}m\overline{v^2},$$

where N denotes the number of molecules in a sample of an "ideal" gas, m the mass of a molecule, v (the absolute value of) the velocity of a molecule, $\overline{v^2}$ the average square velocities of the molecules taken over all of the molecules and over time, and l, w, and h the length, width, and height of the container enclosing the gas.

The question is whether, as Bickle maintains, the derivation of an equation formally analogous to the ideal gas law can be obtained within classical mechanics alone or whether this deduction requires "linking" principles that describe the dependence of certain systemic properties of a macroscopic object (such as $\overline{v^2}$ [the average of the squares of the velocities of all molecules]) on the microscopic properties of its components (e.g., the velocities v of the individual molecules).

The evaluation of the thesis that the CHB model is adequate for the analysis of this particular case of reduction is of some importance. It is a paradigmatic case because of its relative simplicity compared with reductions of chemistry to physics, or of psychology to neurophysiology, which I will consider later. The relative simplicity of its mathematical derivation justifies my consideration of this case as a touchstone: if the conditions imposed by a given model of reduction are too strong for this case to be considered a successful reduction, then the model is inadequate. It is plausible to assume that more complex reductions satisfy these conditions even less.[28]

In an elementary presentation of this reduction, consider an ideal gas of N molecules, each with a mass m, contained in a volume V of a box whose sides define the directions x, y, and z of the Cartesian reference frame. The number of molecules per unit volume is $\rho = N/V$. The fundamental idea behind the reduction is that the pressure of the gas on the walls of the box results from the force exerted by all of the impacts of the molecules on the wall. The aim is to calculate the number of molecules that strike a given surface A during a given period of time Δt and then to multiply it by the force exerted by each of

28 In Krüger's words, it can be assumed that the reduction of thermodynamics to classical mechanics "will mark something like an upper bound to the strength or the completeness one is likely to achieve in reduction in general" (1989, 373).

these impacts on the wall. As a first approximation, we simply assume "that each molecule moves with the same speed, equal to its average speed $\bar{v}$" (Reif 1967, 40).

This simplification contains an innocent part that is relatively easy to abandon as well as a substantial part: the average taken over the speeds of all molecules at a given instant also represents the average over a long time given the dynamics of the molecules. This average only corresponds to a real property of the system when it is in equilibrium. Equilibrium is characterized by the fact that, despite the inevitable changes in the state of motion of the particles, the overall distribution of the speeds of all the molecules is approximately constant (although it undergoes fluctuations around this constant distribution). Once this assumption has been made, the innocent part of the simplification consists of calculating the pressure, not on the basis of the Maxwell-Boltzmann distribution, which indicates the proportion of molecules with a given speed, but on the basis of the overall average speed, taken over all of the molecules. Insofar as this average over the molecules necessarily exists (the number of molecules being finite), this will not change the result of the calculation.[29]

One gets the number of molecules striking the surface *A* over a period of time Δt by assuming that 1/6 of the molecules move approximately in the direction toward *A*, which corresponds to the *x* axis. One in three molecules has a velocity almost parallel to the *x* axis, and one in two of those molecules is moving toward *A*, while the other is moving away from *A*. Next one notes that all of the molecules that hit *A* in an interval Δt must be located in a cylinder (a fictitious construction by the theorist) with face *A* and length $\bar{v}\Delta t$. $\bar{v}\Delta t$ is the distance travelled by a molecule with speed $\bar{v}$ during the time interval Δt. The molecules contained in this cylinder, and only those molecules, will reach *A* during the interval Δt. Based on the assumption that the density of molecules is everywhere $\rho = N/V$, the number of molecules that hit *A* during the interval Δt is therefore $\frac{1}{6}\varrho\bar{v}A\Delta t$. To calculate the force exerted on *A* during a collision of one molecule with *A*, one again uses the assumption that the

29 Richet (2001, 315–16) shows this by calculating the pressure, not simply by multiplying the number of molecules per volume by their average velocity, but by taking the integral of the product of the velocity and the number of molecules with that velocity: $v_x n(v_x)$, where v_x denotes a particular velocity and $n(v_x)$ the number of molecules with that velocity, over all possible velocities, from zero to infinity, using the Maxwell-Boltzmann distribution for the function $n(v_x)$.

gas is in equilibrium. It is assumed that the kinetic energy of such a molecule must remain unchanged during the shock. "This must be true, at least on the average, since the gas is in equilibrium. . . . The *magnitude* of the momentum of the molecule must then, on the average, also remain unchanged" (Reif 1967, 41). Since the force exerted by the wall is equal to the force exerted by the molecule (Newton's third law), and the force exerted by the molecule is equal to its change of the momentum (Newton's second law), it is sufficient to calculate the change of the momentum of a molecule travelling in a perpendicular direction to the wall (the *x* direction), given that the modulus of this momentum remains unchanged: this is the case for a molecule that rebounds after an elastic shock, so that the change in (the *x* component of) its momentum is $-2m\bar{v}$. For the average force (this is the average over both molecules and time) exerted by the molecules on the wall in the *x* direction, one obtains

$$\bar{F} \approx (2m\bar{v})\left(\frac{1}{6}\rho\bar{v}A\right) = \frac{1}{3}\rho m\bar{v}^2 A$$

The average pressure is the average force exerted by the molecules on the wall in the *x* direction, per unit area of the wall:

$$(1)\quad \bar{p} \approx (2m\bar{v})\left(\frac{1}{6}\rho\bar{v}\right) = \frac{1}{3}\rho m\bar{v}^2$$

The crucial question is this: does this calculation allow us to deduce, from the concepts and principles of classical mechanics alone, which is the reductive microtheory describing the behaviour of microscopic particles, an expression that describes a macroproperty of the macrosystem of which these molecules are the components? This is the fundamental condition of the CHB model, which applies, according to its proponents, to the case of the reduction of thermodynamics: no linking principle is necessary. The first part of the reduction consists of deducing, using only the concepts and principles of the reducing theory, an analogous image of the reduced theory. This analogous image is supposed to have two properties: it is entirely formulated in the language of T_B (mechanics), and it nevertheless describes global (or systemic) properties in the domain of the reduced theory $T_R{}^*$ (thermodynamics). The second part of this statement is true but not the first part. It is true that *v* is a term belonging to the vocabulary of the reduced theory. It is reasonable to argue that this is also the case for the average speed of molecules at a given instant. In favour of this hypothesis, it can be argued that this average speed is always the speed of a microscopic object, albeit a virtual object: there is as much reason to say this as there is to say that the centre of gravity of a system of massive bodies, albeit virtual, is a mechanical object, at the same level as the massive bodies that make up the system.

Now the velocity $\bar{v}$ in (1) is an average obtained by neglecting the fluctuations of individual molecules. It is only on this condition that it is possible to conceive of the expression on the right in (1) as designating a macroscopic property of the gas. One can only reduce the temperature of the gas to the kinetic energy of the molecules if one defines this kinetic energy on the basis of such an average.

$e_{kin} = \frac{1}{2}mv^2$ (where e_{kin} denotes the kinetic energy of a molecule, v its velocity, and m its mass)

is a microscopic property that can be attributed to the molecular components of the gas. But the quantity that forms the basis for reducing temperature as a macroscopic quantity is

$$\overline{e_{\text{kin}}} = E_{kin} = \frac{1}{2}m\overline{v^2}$$

To be able to consider that an expression refers to a macroscopic property of the gas, it is necessary to assume that the averages over time correspond to real properties; in short, it is necessary to assume that the system is in equilibrium. "The temperature and pressure of a gas have only a statistical significance. An isolated atom has no temperature and pressure. . . . Temperature and pressure can be defined only when the number of atoms is large enough that their values are time independent" (Richet 2001, 316).

The conceptual transition from the microscopic mechanical domain to the macroscopic thermodynamic domain therefore corresponds to the transition from a system made up of a set of molecules whose average speed and kinetic energy over the macroscopic sample and over time can be calculated, to a single object with the stable global properties p and T. In order to describe and explain the behaviour of a macrosystem in thermodynamic terms, it must be assumed that it actually possesses these properties. For the averages of microscopic quantities over time to correspond to real properties of the system, the system must be in equilibrium. If this is not the case, the average $\overline{e_{kin}}$ over time does not give rise to a temperature, and the average $\bar{v}$ over time does not give rise to a pressure: the system does not possess these latter properties, and it does not obey the laws that define them, such as the "zeroth law of thermodynamics," which defines temperature in terms of equilibrium.

One might suspect that the need to resort to the equilibrium hypothesis is merely an artifact of the simplified presentation of reduction in textbooks. This is not the case. In the presentation of the reduction of thermodynamics by Gibbs (1902) — which still constitutes the most important theoretical model — the conceptual leap between the consideration of a system as composed of a set of microcomponents and the consideration of a thermodynamic macrosystem is clearly apparent. It is impossible to know the exact state of the molecules. Even at equilibrium, the real values of the mean pressure at a given moment, and of the average kinetic energy $\overline{e_{kin}}$, taken over all of the molecules at a given moment, are not strictly identical to their mean values taken over time but fluctuate around this mean value. Yet the macroscopic properties T and p by definition are properties that characterize systems at equilibrium: by definition, they are independent of time. In Gibbs's approach, the values of (macroscopic) thermodynamic quantities are calculated from the fiction of the "set" of all possible microsystems given the macroscopic constraints imposed. This construction ensures that the values of these quantities remain stable over time. This means that thermodynamic properties, such as temperature, are not calculated directly from the state of the particular underlying microscopic system. They are calculated from the fiction of all the systems obeying the same constraints. Insofar as the system is in equilibrium, it really does have a macroscopic temperature and pressure. These properties are not fictitious. They are real properties whose conception is irreducible to the framework of microscopic mechanics alone. The conceptual transition corresponding to Gibbs's derivation of these quantities from the description of the system in terms of the microproperties of its components cannot be accomplished with the conceptual resources of the reducing theory alone. The use of the Gibbsian concept of an "ensemble" of systems (or some other concept that cannot be reduced to the conceptual apparatus of microscopic mechanics) is indispensable.[30]

30 Nagel is aware of the indispensable nature of statistical premises, themselves irreducible to mechanics, in the reduction of thermodynamics: "It is one thing to say that thermodynamics is reducible to mechanics when the latter counts among its recognized postulates assumptions (including statistical ones) about molecules and their modes of action; it is quite a different thing to

Regarding pressure, Sklar expresses this point as follows:

> There is, for a particular sample of gas at equilibrium, the actual momentum transferred by the molecules impinging on a wall of the box in a short time, and there is its average value per unit area per unit of that time. On the other hand, there is the quantity calculated for an ensemble of similarly constituted systems. . . . Whereas the former sort of pressure, the feature of the individual system, will be expected to fluctuate, the latter kinds of ensemble quantities, quantities defined by the macroscopic characterization of the system and the chosen probability distribution over the ensemble, will, of course, not. Here, fluctuation will show up as assimiliated into the ensemble description by the calculation of averages or most probable values of quantities, but the averages themselves are not the sort of things to fluctuate. (1993, 349–50)

The "orthodox" procedure for reducing thermodynamic quantities to mechanics involves Gibbs's concept of ensemble. Of course, the fact that the derivation of thermodynamic concepts via the ensemble concept is not purely mechanical does not mean that there is no other way of deriving them that does not require recourse to concepts that do not belong to mechanics.[31] However, other attempts to reduce thermodynamics have encountered the same difficulty. For example, it has been suggested that the concept of probability should be avoided when moving from the description of a system in microscopic terms to its description in macroscopic terms, by deducing macroscopic laws from mechanical laws and initial conditions; it is plausible to think that the non-existence of macroscopic systems that break macroscopic laws while obeying microscopic mechanical laws (e.g., an isolated system whose entropy decreases "spontaneously" without being compensated by an increase in entropy in another system) can be explained by the fact that

claim that thermodynamics is reducible to a science of mechanics that does not countenance such assumptions" (1961, 362). According to my analysis, the first claim is justified but not the second claim. However, the reduction would only satisfy CHB's conditions if the second statement were true.

31 Krüger (1989) briefly presents three other approaches. None of them uses only mechanical conceptual resources.

this would require exceptional initial conditions. Clearly, we cannot, in practical terms, specify such initial conditions since we can neither observe nor describe the state of motion of 10^{23} particles at a given moment.[32] But let us put aside this difficulty; it might be only an epistemic one that does not affect the ontological aspect of the question. Now, even if we ignore the problem of the number of factors that form part of the initial conditions, there is no reason to think that the initial conditions that characterize systems obeying macroscopic laws can be rendered in microscopic terms. On the contrary, it seems to be plausible that the only commonality of these conditions is the *macroscopic* property of characterizing systems that obey macroscopic laws.[33] The specification of initial conditions, if it were practicable, is therefore possible only by using macroscopic concepts and vocabulary. Consequently, a reduction of thermodynamics along this path would not be able to deduce it from the conceptual resources of microscopic mechanics alone.

It can be concluded provisionally that in the present state of science the CHB model does not apply to the reduction of thermodynamics to mechanics. Its principles cannot be derived from mechanical resources alone. As Krüger concludes, "notions like equilibrium and temperature (and thereby entropy) must be given physical meaning on a basis of more than just mechanics" (1989, 382).[34]

---

32 There are approximately 10^{23} atoms or molecules in 1g of matter. More precisely, the "Avogadro number," defined as the number of carbon atoms in 12g of the 12 of carbon, is approximately equal to $\mathbf{6 \cdot 10^{23}}$.

33 "There is, as far as I am aware, no indication that there is a non-trivial or non-question-begging property in the language of mechanics that would be common to all microstates which behave normally, but absent from all those which do not" (Krüger 1989, 379).

34 This is independent of the interpretation of the concept of ensemble. Gibbs himself interprets it as an expression of our ignorance of the detailed microscopic state: "The laws of thermodynamics . . . express the laws of mechanics for these systems, as they appear to beings who do not possess sufficient fineness of perception to be able to appreciate quantities of the order of magnitude of those belonging to the individual particles, and who cannot repeat their experiments often enough to obtain other than the most probable results" (1902, vii–viii; cited by Krüger 1989, 380). For Gibbs, we have to construct ensembles because we do not have access to the individual properties of all the microscopic particles that make up a macroscopic system. Einstein, who developed the ensemble approach independently of Gibbs, interprets ensembles in a different way. For Einstein, the ensemble describes the distribution of energies in a collection of systems in contact with a "heat bath" (i.e., an infinitely large reservoir of heat that has a fixed temperature) (1902, para. 5; 1903, paras. 3–4; Krüger 1989, 382). In Einstein's interpretation, a fictitious infinite ensemble serves as the basis for assigning thermodynamic quantities to a real individual system. Irrespective of the

One of the attractions of the CHB model is its promise to account for reduction without having to postulate linking principles that give rise to the suspicion of making the reduction obscure. As long as these binding principles themselves are not reduced (by derivation from T_B), the higher-level theory is only incompletely reduced but remains partly mysterious. In Bickle's words, "one advantage of the H-C [Hooker-Churchland] account is that it avoids having to specify the logical status of cross-theoretic identity statements" (1992, 223). It avoids this problem because, "if the deductive part of a reduction has no gap to bridge between the language or the ontology of premise and conclusion, then the nonexistence of lawlike connections between reduced and reducing concepts or kinds is of no consequence" (Bickle 1998, 108). We have seen that the CHB model, far from circumventing the problem, puts forward a hypothesis to solve it: it is the hypothesis that the relevant "identity statements" can be derived within T_B (or *approximately* derived, insofar as T_R^*, an approximation of T_R, is strictly derived). However, I have shown that this assumption is false in the case of the reduction of thermodynamics.

Bickle's thesis that the CHB model of reduction does not need linking hypotheses is crucial in his defence of reductionism against various anti-reductionist arguments. It is important to show that Bickle's failure to refute these arguments does not refute reductionism. The "synthetic model of reduction," which I will introduce in the next section, makes it possible to answer them without the thesis (which, as we have seen, is mistaken) according to which T_R^* can be derived from T_B without linking statements.

1. Davidson (1970) justifies the autonomy of psychology by the absence of strict psychophysical laws. This argument presupposes that the reduction of psychology to neurophysiology requires the discovery of psychophysical laws that can play the role of the linking principles in Nagel's model. However, Bickle claims that "the impossibility of psychophysical laws is irrelevant to the new thesis of mind-brain reductionism and the novel account of intertheoretic reduction underwriting it" "since an H-C [Hooker-Churchland] reduction nowhere requires bridge laws" (1992, 218, 224). This defence of reductionism against Davidson's argument invites two objections.

First, the absence of linking principles in the CHB model does not stand up to the test of the analysis of a paradigmatic case of reduction. This analysis

interpretation chosen, the need to make use of a fictitious infinite ensemble shows the inadequacy of strictly mechanical concepts for deriving thermodynamic properties and their laws.

shows that at least some reductions require linking principles. The analysis of the reduction of long-term memory will show (in section 7) that it involves laws of composition that play the role of linking principles. Davidson's thesis, according to which the difference between the conditions of attribution of physical and psychological states precludes the existence of psychophysical laws, cannot therefore be true in general. It is still possible that this thesis is true of a (very important) part of our psychological states: it is possible that there are no linking principles that directly connect *intentional* states, such as propositional attitudes of believing and desiring, to states of the brain. However, such intentional states may be related nomologically to other mental states that *are* reducible to brain states.

Second, as Endicott has pointed out, the CHB model itself contains linking principles between T_R^* and T_R: in the case of relatively "retentive" reductions, where the corrections that T_R^* makes to T_R are modest, the reduction justifies identities between objects and properties described by the theories T_R^* and T_R (see Churchland 1979, 83; 1985, 11). Furthermore, "property identity guarantees nomic coextension. So bridge laws exist within the new-wave account" (Endicott 1998, 68): that is, in the CHB model of reduction.

2. Bickle sees scientific theories as sets of models rather than sets of statements.[35] In logic, a "model" of a statement (or set of statements) is an interpretation in which the statement (or set of statements) is true. An "interpretation," in the logical sense of the term, associates an object with each singular expression and a set of objects with each predicate. The "semantic conception" of scientific theories consists of conceiving of theories not as sets of statements but as sets of models: the structured sets of objects that make the theory true. According to Bickle, adopting the semantic conception makes it possible to analyze the relationship between reduced and reductive theory without resorting to linking principles. Although this assertion is undeniably true in a literal sense, it seems to be rather superficial. Indeed, in the case of "homogeneous ORLs" (Bickle 1998, 78), where ORL stands for ontological reductive link, the basic sets of the models of T_R constitute a subset of the basic sets of the models of T_B. For example, the point masses of classical collision mechanics (T_R) are identical to the point masses of Newtonian

35 In other words, Bickle adopts the "semantic conception" of scientific theories, an important alternative to the traditional "syntactic conception," according to which theories are sets of statements.

particle mechanics (T_B).[36] Even if, at a fundamental level, theories are sets of models and not sets of statements, they *imply* statements. The assertion (part of the requirements of the CHB model) that the entities designated by the statements of T_R are identical to the entities designated by the statements of T_B can be presented as a consequence of inclusive relations of the basic sets of the respective models of T_R and T_B; this does not prevent the fact that, as soon as such an identity is expressed in a statement, it is a linking statement. In other words, the CHB model still contains statements of intertheoretical connection, even if they occupy the place of consequences and not that of fundamental postulates.[37]

The attempt to identify interlevel reduction with corrective intralevel reduction therefore fails, at least in this paradigmatic case, for the reason that the deduction of T_R^* from T_B cannot be conceived of as intralevel within T_B. Another observation diminishes the plausibility of this assimilation. The corrective modification of the reduced theory, although it is a regular effect of reduction, is not always the unique aim of interlevel reduction.[38] Another important aim is explanatory unification: a reduction aims to show what macroscopic properties such as temperature *consist of*. It is perfectly possible for this goal to be achieved by an interlevel reduction that justifies the high-level theory as it is, whereas the only reason for an intralevel reduction is to improve the reduced theory. Interlevel reduction provides information about the detailed nature of the reduced properties, without necessarily showing that the reduced theory is wrong: reduction justifies the notion of temperature as

36 See Balzer, Moulines, and Sneed (1987, 255–67); Bickle (1998, 78).

37 Schaffner had already shown that the "semantic" conception of reduction first proposed by Suppes (1967), conceiving of it as a relation of isomorphism between models of theories rather than between their statements, "is a special case of a Nagel type of reduction" (1993, 430; Schaffner 1967, 145). The isomorphism of models is not sufficient for a reduction because theories can relate to domains that have isomorphic models without being connected. Schaffner (1967, 145) mentions heat theory and hydrodynamics, which share their formal structures, without one being reducible to the other. For this reason, Bickle adds the condition of the existence of ORLs. With this condition, the semantic conception of reduction becomes equivalent to the syntactic conception in terms of linking statements. Furthermore, Endicott (1998, 71) points out that the only innovative element of the CHB model compared with Schaffner's model — namely, the requirement that T_R^* be inferred from the conceptual resources of T_B alone, cannot be expressed in the semantic conception, because the notion of inference in accordance with rules makes sense only in the context of a conception in which theories are expressed in statements.

38 "Unlike the evolutionary intralevel cases, the reduced theory in interlevel situations does not stand in need of technical correction in every case" (McCauley 1996, 30).

a stable (time-independent) property of macroscopic systems at equilibrium, but it also provides the information that this stable property emerges against a background of microproperties subject to fluctuations and themselves therefore stable only on average.

The lesson to be learned from the analysis of the reduction of thermodynamics is that macroscopic thermodynamic properties cannot be *identified* with microscopic properties of their components. It is not enough to assume that gases are composed entirely of molecules to be able to reconstruct the macroscopic properties of gases from the microscopic properties of their components. "A too naive application of the notion of identificatory reduction would be misleading, because 'temperature' in many of its uses in statistical mechanics refers not to an instanced property of a particular system sample at a time, but, rather, to some feature of a probability distribution over systems of a specified type" (Sklar 1993, 352).[39]

6. The Synthetic Model of Reduction

We have seen that the paradigmatic reduction of thermodynamics to classical mechanics is possible only on the basis of concepts and principles that go beyond those available in the reducing theory. In what follows, I propose a "synthetic" model that takes this into account. This model, sketched in Figure 1.3, retains elements of each of Nagel's model and the CHB model.

39 In Chapter 4, we will see that we can characterize the relationship between a macroproperty such as temperature and the microproperty that allows its reduction more adequately using the concept of emergence. The properties of a macroscopic object composed of microscopic parts are determined by laws of composition. When the law of composition is non-linear, macroproperties can emerge that are qualitatively different from the microproperties that determine them.

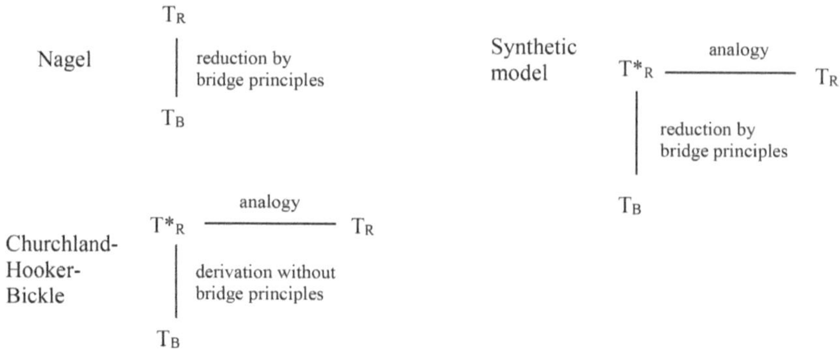

Figure 1.3 Three reduction models, differing with respect to bridge principles and historical change.In Nagel's model, the old theory T_R is deduced from the new theory T_B by means of linking principles. In the CHB (Churchland-Hooker-Bickle) model, this deduction is carried out without any use of linking principles, using only the conceptual resources of T_B, and it leads to a new theory $T_R{}^*$, analogous to the reduced theory T_R. $T_R{}^*$ is at the same theoretical level as T_R but is superior to it. In the synthetic model (inspired by Schaffner), the reduction of T_R involves deducing a theory $T_R{}^*$ that corrects T_B, starting from T_B *using linking principles.*

The synthetic model retains Nagel's thesis that the linking principles (or bridge laws) cannot be deduced without prior knowledge of the phenomena and laws of the level of the reduced theory ($T_R{}^*$ and T_R). The derivation of $T_R{}^*$ from T_B presupposes the a posteriori discovery of interaction laws as well as concepts and laws specific to the transition between levels. My model also retains the observation of the CHB model that the result of such a reductive deduction does not in general coincide exactly with the old theory T_R but that it yields corrections. In short, the synthetic model takes account of the fact that reductions generally lead to a correction of the reduced theories without requiring the reduced theory to be absorbed by the reducing theory. In general, a reduction consists of the discovery of a nomological explanation of structures and laws governing the evolution of these structures, which uses conceptual resources from the reducing theory T_B and the reduced theory T_R. Indeed, as we saw in the previous section, the reduction of thermodynamic laws uses, in addition to the principles of the reducing theory, statistical principles foreign to microscopic theory, in particular Gibbs's notion of an ensemble.

Schaffner's model does not explicitly require the derivation of $T_R{}^*$ to use only the conceptual resources of T_B. In this respect, it resembles the synthetic model that provides the possibility of deriving T_R* using conceptual resources specific to T_R. Schaffner takes up the Nagelian distinction between "homogeneous" and "heterogeneous" reductions. In the case of the former, "all the primitive terms . . . appearing in the *corrected* secondary theory T_R* appear in the primary theory" (1967, 144; variables modified), whereas in the case of the latter the primitive terms of T_R* are associated only with the terms of T_B, where these associations are subject to a certain number of conditions. But these conditions do not limit the conceptual resources that can be used in the construction of T_R* (i.e., resources from T_R as well as those from T_B).[40] Schaffner's GRR model ("general model of reduction-replacement") requires only the existence of a function that associates the predicates of T_B (possibly corrected to T_B*[41]) with the predicates of T_R*. The last condition can be satisfied if the properties of T_R* are determined, as a function of the properties of T_B*, by non-causal laws[42] of composition, according to my synthetic model.

This is an important difference, and Schaffner's model partially escapes my criticism of the CHB thesis, according to which the resources of T_B alone are sufficient to derive T_R*. Nevertheless, the conditions that Schaffner imposes on reduction are too strong: the paradigmatic case of the reduction of thermodynamics does not satisfy them because Schaffner imposes that, even in heterogeneous reductions, the terms of T_R* must have the same reference as the expressions constructed in T_B associated with them during the reduction. The association of T_R* terms referring to individuals, with expressions formed in T_B, takes the form of a "one-to-one correspondence representing synthetic identity between individuals . . . in T_B and T_R*," and "the primitive predicates of T_R* . . . are . . . associated with an open statement in T_B," so that the reduction function that accomplishes this association is "in general . . . interpretable as expressing referential identity" (Schaffner 1967, 144; variables

40 Endicott correctly observes that "Schaffner does not *require* that T_R* . . . be constructed out of the [conceptual] resources of a higher-level T_R" (1998, 58n14). But Schaffner's model does not require T_R* either to be constructed merely from the conceptual resources of T_B.

41 The GRR model takes account of the fact that fruitful reductions often also lead to a modification of the reducing theory. The reduction consists of deducing a corrected theory T_R* from a corrected reducing theory T_B*. See Schaffner (1993, 427–29).

42 The assimilation of the relation of simultaneous non-causal determination to non-simultaneous causality is a major source of confusion in the philosophy of reduction and mental causation. I will explain in Chapter 5 that this confusion plays a key role in Kim's argument against the causal efficacy of macroscopic properties.

modified).[43] However, this thesis is questionable at least with regard to predicates. We saw in the previous section that, in the case of the reduction of the thermodynamic properties of pressure *p* and temperature *T*, there is no expression in the vocabulary of T_B that shares their reference.

7. The Reduction of Cognitive Phenomena by Neurophysiology: Elimination or Co-Evolution?

According to my working hypothesis, reduction is one of the forms of unification of scientific knowledge. Therefore, it will be instructive to compare my analysis of the reduction of thermodynamic quantities in the physical sciences with a case of reduction in the cognitive sciences.

Recent research in neuroscience has uncovered the neural bases of several basic forms of learning: sensitization, habituation, and classical conditioning. Sensitization is a form of learning in which an animal learns to react differently to a stimulus: repetition of the stimulus leads to reinforcement of the behavioural response. Habituation, conversely, reduces the strength of the animal's reaction to a stimulus repeatedly presented. These forms of learning are non-associative in that they involve a single stimulus and a single response. In contrast, classical conditioning (CC) is a form of associative learning in which an animal learns to react with a behavioural response R to a stimulus (CS for *conditioned stimulus*) that is neutral before learning (i.e., that does not provoke any behavioural response), by associating this CS with another stimulus, known as the *unconditioned stimulus* (US), which provoked response R before the learning session. It can be innate that the US triggers R.

The reduction of these elementary forms of learning is well advanced in science.[44] The results presented by Hawkins and Kandel (1984) enable us to understand the neurophysiological mechanisms by which conditioning and

---

43 True, the requirement that the connection between the predicates of the theories T_R* and T_B (or T_B*) must be established by a hypothesis of "synthetic identity" does not appear in the formal conditions of the GRR model (Schaffner 1993, 429). Nevertheless, Schaffner keeps maintaining that "reduction functions" establish that entities designated by predicates of T_R* are identical to entities designated by predicates of T_B*. "Reduction functions link entities and predicates of reduced and reducing extensionally via an imputed relation of synthetic identity" (Schaffner 1993, 440).

44 There are other cases of detailed neurobiological reductions of cognitive capacities, such as the phenomenon of colour opposition, the basis of which has been discovered in neurons known as *colour opponent cells*. See Hardin (1988); Zeki (1993); Gold and Stoljar (1999). I will examine this reduction in Chapter 4.

learning occur: in a situation in which the animal is confronted with the CS, it reacts with the learned behaviour R.

The importance of reducing these forms of learning would be even greater if it turned out that sensitization, habituation, and classical conditioning form a "learning alphabet" whose combination makes it possible to explain more complex forms of learning. Based on the discovery that the microscopic processes underlying classical conditioning (a type of associative learning) are variants of the processes underlying sensitization (a type of non-associative learning), Kandel makes the hypothesis that "more complex forms of learning can be built up from the molecular components of simpler forms. By this means a variety of distinct forms of behavioral modifications could be achieved by a small set of molecular mechanisms" (1995, 680; see also Hawkins and Kandel 1984). This perspective allows us to answer the objection that the reduction of elementary learning is of very limited significance for psychology, insofar as this reduction has been achieved only in the case of a rather primitive creature, the species *Aplysia californica*: a naked-bodied gastropod mollusc of the genus *Aplysia* (some of whose species are also known as "sea hare") whose cognitive capacities are viewed as rudimentary. A great deal of research on the physiological basis of learning has been carried out using the species *Aplysia californica*, which lends itself to this type of study because of the simplicity of its neural system and the large size of its neurons. However, the primitive forms of learning studied in *Aplysia* are shared by much more complex animals and even by humans. We therefore accept the widespread view that Hawkins and Kandel's discovery is a paradigmatic case for cognitive neuroscience, which aims to discover the neurophysiology underlying cognitive abilities.[45]

Let us start with a form of sensitization that has been studied at the cellular level in *Aplysia*.[46] *Aplysia* has an innate reflex that consists of retracting the siphon inside the parapod and the gill in the mantle (R) as a consequence of a threatening stimulus (US: a tap on the mantle or siphon).[47] This gill-withdrawal reflex is present before sensitization. Sensitization is a process that

45 See Churchland (1986, 369); Bickle (1998); Gold and Stoljar (1999).

46 See Hawkins and Kandel (1984, 377–78); Hall (1992, 474 ff.); Kandel (1995, 671–76).

47 A stimulus is called unconditional (US) if it triggers a behavioural response without any prior learning, as happens with innate reflexes. A stimulus is said to be conditional for a given response R if the CS triggers R only after learning by classical conditioning, whereby the animal learns to associate the CS with a US and therefore with the response R triggered by the US.

leads to the reinforcement of this reflex. It is triggered by a noxious stimulus ($US_2 = N$) to another part of the body, in this case a shock to the tail or head. In psychological terms, the state provoked by *N* can be interpreted as a state of alert that reinforces all defensive behaviours (Hawkins and Kandel 1984, 381).

Here is a simplified explanation of sensitization. The neurons that transmit information about *N* make synapses on facilitator interneurons. These interneurons make synapses on the axon ending of the sensory neuron that transmits information about the US, precisely at the point where this axon has a synapse with the motor neuron responsible for triggering R. The mechanism of sensitization is presynaptic facilitation: the US-R synaptic connection is modified by the interneuron so that R responds more strongly to the US. This is achieved by a modification of the dispositions of the molecular parts of this synapse.[48] Sensitization is reduced in two stages. In the first stage, the cognitive process is reduced to a neurophysiological process of synapse modification. In the second stage, the central stage of the neuronal mechanism is in turn reduced to molecular processes.

Several molecular changes occur in parallel. Stimulation of the interneuron (stimulated by *N*) causes, via a biochemical mechanism, the closure of a number of K^+ channels in the US-R presynaptic axonal ending. When an action potential triggered by a US arrives at the axonal ending, open K^+ channels tend to bring the potential difference across the axonal membrane back to its equilibrium value, whereas closed K^+ channels increase the depolarization that determines the strength of the action potential. Stimulation by *N* thus leads to "changing the conformation of the channel and decreasing the K^+ current," which "prolongs the action potential, increases the influx of Ca^{2+}, and thus augments transmitter release" (Kandel 1995, 673) into the US-R intersynaptic channel.

The reductive explanation of classical conditioning also proceeds by highlighting microscopic changes that result in a modified state of the neurons involved, which can be characterized alternatively in categorical terms or in dispositional terms. Classical conditioning leads to establishing or reinforcing an animal's disposition to react by R to the perception of a CS to which it reacted little or not at all before learning. For this conditioning to take place,

48 In Chapter 3, we will see that cognitive properties can be described both categorically, by describing them in themselves, in abstraction from their role or function, and dispositionally, by identifying them through their causes and effects.

the sensory neurons originating from the CS must be stimulated in a precise temporal interval that precedes the stimulation of the same axonal bulb by the US.

The first stimulation by the CS puts the presynaptic axonal bulb in a state of greater receptivity to successive stimulation by the unconditional stimulus. The molecular basis of this greater receptivity consists of a change in the conformation of adenyl cyclase, a molecule involved in the mechanism leading from stimulation by an action potential to release of the transmitter into the intersynaptic cleft.[49]

Each reduction of a cognitive capacity by the discovery of an underlying microscopic mechanism relies on laws of interaction between the microscopic parts of the system. The overall determination resulting from these laws can be considered to be based on a single "law of composition" (Broad 1925, 63) specific to this type of system. This law determines, in a non-causal way, that the complex system possesses the overall property *M* because its parts (the various ion channels, transcription activators, etc.) possess certain properties: the property of a fraction of the K^+ channels to be closed (P_1), the property of the transcription activators to be phosphorylated (P_2), and so on. The reduction shows how the global cognitive property *M* nomologically depends[50] on microscopic properties $P_1 \ldots P_n$ of parts of the system. In what follows, I will sketch another example of reductionist explanation in cognitive neuroscience before using these examples to evaluate the synthetic model of reduction introduced above.

The discovery of the mechanism underlying the acquisition of long-term memory is another example of a well-advanced psychophysiological reduction. Experimental research in cognitive science has shown that the transformation of short-term memory into long-term memory requires (most of the time) repetition of the stimulus and can be prevented by "retrograde interference": that is "distractions introduced after the initial items have been learned and stored in short-term memory" (Bickle 2003, 47).

49 The conformational change of the molecule "enhances its ability to synthesize cAMP in response to serotonin released in the US pathway" (Kandel 1995, 679). cAMP is short for cyclic adenosine monophosphate, one of the molecules involved in the mechanism of long-term memory consolidation. See Kandel (2000, 1254).

50 It is equivalent to say that *M* depends nomologically on $P_1 \ldots P_n$ and to say that $P_1 \ldots P_n$ determine nomologically *M*.

Several molecular mechanisms underlie the development of long-term memory. The first mechanism corresponding to early long-term potentiation (E-LTP) gives rise to a synaptic modification that persists for approximately one to three hours (see Bickle 2003, 63–67). Through a cascade of biochemical interactions involving numerous molecules, the strong depolarization of the post-synaptic membrane leads — through the reception of molecules of the neurotransmitter glutamate from the synaptic cleft and after numerous intermediate steps involving other channel proteins in the membrane as well as cAMP molecules (secondary messengers) — to a change in the conformation of two species of receptor molecules, called AMPA and NMDA. The consequence of this conformational change is that the channels with which these molecules are associated remain in an "open" state. In this state, the conductivity of the AMPA receptor for Na^+ ions, for example, is almost tripled. This means that, if a new stimulation reaches the post-synaptic neuron while it is in the long-term potentiated state, it will produce an enhanced post-synaptic excitatory potential (EPSP), increasing the likelihood that the overall depolarization at the axon neck of the post-synaptic cell will be strong enough for it to send out an action potential in response to stimulation by the pre-synaptic cell.

The second phase of long-term potentiation (L-LTP) is triggered by repeated stimulation of the post-synaptic cell. It leads to a much longer-lasting change in the structure of the post-synaptic neuron. L-LTP essentially involves gene expression. It is triggered by a product of the post-synaptic causal chain, the catalytic molecules PKA, which migrate to the nucleus of the cell, where they trigger expression of the *uch* gene (see Bickle 2003, 67–71). This gene transcribes the protein hydrolase ubiquitin, which triggers the transcription of other proteins that cause the growth of new dendritic spines and hence the formation of new synapses. The end result of this mechanism underlying long-term memory is an increase in the number of synapses between two neurons. This, in turn, increases the likelihood that the post-synaptic neuron will send out an action potential when stimulated, however weakly, during the L-LTP period, which can last for days or weeks.

The discovery of this molecular mechanism underlying the formation of long-term memory provides a reductive explanation of a number of properties of long-term memory first highlighted at the cognitive level. I will mention just two of them.

First, research in experimental psychology conducted at the end of the nineteenth century established that there is a linear dependence between the number of times that a stimulus is repeated during conditioning and the length of time that it is retained in long-term memory.[51] The description of the underlying microscopic processes makes it possible to explain this dependence, insofar as the brain event triggered by the repetition of the stimulus produces the state underlying memory consolidation. Stimulus repetition leads to enhanced activation in presynaptic axons, triggering the cascade of biochemical events described above, which gives rise to structural changes in synapse configuration. This explains the neuron's increased disposition to respond to similar stimuli.

Second, it has also been known since the end of the nineteenth century that brain trauma suffered after the initial phase of learning can prevent the fixation of memories related to the period immediately preceding the shock. Experimental work on this phenomenon, known as "retrograde interference," shows in particular that, if an animal is given an electric shock between twenty seconds and fifteen minutes after having undergone an experience whose memory is stored in short-term memory, that memory will not get fixed in long-term memory.[52] Furthermore, the retrograde amnesia that concussion victims suffer is explained at the neuronal level by the fact that the trauma interrupts one of the biochemical stages of the process leading to L-LTP.

We have seen with the example of thermodynamics that its reduction to classical physics does not make the reduced theory superfluous. First, the discovery of the laws that determine macroscopic properties and processes presupposes prior knowledge of macroscopic phenomena and laws. Second, the deduction of macroscopic phenomena necessarily involves concepts and principles that cannot be deduced a priori from the principles and laws governing microscopic phenomena alone. The reductive explanation, therefore, leads to a unification of knowledge and the enrichment of both theories rather than the elimination of the reduced theory.

---

51 The work of Ebbinghaus is presented in Squire and Kandel (1999, 130–32) and Bickle (2003, 47). However, repetition of the stimulus is not necessary. Depending on the biological species and the object of learning, certain experiences can be sufficient, without ever being repeated, to induce a stable long-term memory.

52 After Ebbinghaus and Müller and Pilzecker at the end of the nineteenth century, this research was taken up again by Duncan (1949). See Bickle (2003, 112).

Similarly, neurophysiological reduction leads to a deeper understanding of cognitive phenomena, such as learning, but does not render their psychological descriptions superfluous. Specifically cognitive concepts are indispensable for understanding certain properties of learning by conditioning. Consequently, a cognitive concept such as having learned to react with a behavioural response to the perception of the conditioned stimulus is neither eliminated when the underlying microprocesses have been discovered nor identified with any concept applying to those microprocesses. The cognitive concepts describing learning at the cognitive level remain indispensable insofar as they are necessary for the very description of the underlying microscopic mechanism. According to Rescorla (1988), such a mechanism cannot be understood in terms of a "low-level mechanical process in which the control over a response is passed from one stimulus to another" (152). Indeed, the mechanism underlying learning becomes comprehensible only insofar as we appeal to the notion of information (see Gold and Stoljar 1999, 31). Rescorla shows that there are two ways of conceiving of learning. First is the "reflex tradition in which Pavlov worked and within which many early behaviourists thought" (152). In this research tradition, conditioning is interpreted "as a kind of low-level mechanical process" (152). Second is "the associative tradition," which "sees conditioning as the learning that results from exposure to relations among events in the environment," where "the information that one stimulus gives about another" is crucial (152). Rescorla presents the transition from the old to the new theory of learning by conditioning as the replacement of a theory whose concepts are located at the physical level by an authentically psychological theory, constructed using the concept of information. This interpretation seems to be too simple: it is also a matter of replacing a crude theory with a more sophisticated one. However, the crucial point for my analysis of reductive explanation is the fact that the improvement of the theory of learning is the result not of the discovery of neurophysiological mechanisms but of the use of the cognitive concept of information.

As theories of the mechanism of learning, both theories can be expressed in cognitive vocabulary. According to the crude theory, as expressed in textbooks from the 1980s, conditioning is "a form of learning in which a neutral stimulus, when paired repeatedly with an unconditioned stimulus, eventually comes to evoke the original response" (Gardner 1982, 594, cited by Rescorla 1988, 151). For conditioning to be effective, the presentation of the CS and the US must follow a precise protocol. The CS must be presented within a well-defined interval before the US is presented, an interval known as the interstimulus interval (ISI); optimal classical conditioning requires an ISI of half a second (see Hawkins and Kandel 1984, 379). The more sophisticated theory presented by Rescorla shows that the concept of *contiguity* between the unconditioned stimulus and the conditioned stimulus is too crude: contiguity is neither necessary nor sufficient for conditioning to occur.

The fact that it is not *sufficient* is demonstrated by Rescorla's experiments on rats. In these experiments, rats were exposed to two salient events: a tone lasting two minutes and a weak electric shock applied to the grid on which they were standing. Two experimental protocols were compared. In the first, there was no temporal correlation between the two events, so tones contained no information about shocks. Some shocks occurred in the presence of a tone, others in its absence. In the second, the rats were exposed to tones that had the same distribution as in the first protocol; however, they were exposed to shocks only in the presence of tones, and there were no shocks in the absence of tones. Both protocols satisfied the condition of contiguity between the US (the shock) and the CS (the tone), but only the rats that followed the second experimental protocol developed an association between the tone and the shock. Rescorla explains this result by the fact that, in the second protocol but not in the first, the tone contained information about the shock. In the first protocol, there was as much probability of a shock when there was a tone as when there was no tone; in the second protocol, shocks occurred only during tones. The two learning situations "share the same contiguity of the tone [the CS] with the US, but they differ in the amount of information that the tone gives about the US" (Rescorla 1988, 152). In the first group, the presence of the CS informed the animal of the presence of the US, which explains the creation of a conditioned response to the CS previously adequate only for the US. Conversely, in the second group, the CS contained no information about the US because there was as much US in the absence of CS as in its presence, which explains why the animal did not develop a conditioned response to the CS.

Contiguity is not *necessary* for learning either. In a variant of the experiment described above, starting from the protocol in which the US was present simultaneously with the CS and in the absence of the CS, all US contiguous with CS were removed. Since there was no more contiguity between CS and US, the simple theory of learning by contiguity predicts that no learning will occur. However, what the animals in fact learn in this situation is that the

CS is a reliable indicator of *the absence of* the US. They learn by developing a conditioning in which the CS acts as a conditional inhibitor (see Rescorla 1988, 153).

Rescorla takes the old Pavlovian theory of the reflex to be a "mechanistic" theory that draws its conceptual resources exclusively from the neuronal level and the cognitivist theory of learning to be a theory that makes an irreducible use of the concept of information. Now this reconstruction is as questionable as Bickle's (2003) eliminativism, which claims that the discovery of the underlying biochemical processes renders the use of psychological concepts superfluous. It seems to be more appropriate to interpret the difference between behaviourist and cognitivist theories of learning by conditioning in terms of how fine a distinction they draw. In fact, each of these theories (or each of these variants of the theory) has a corresponding reducing theory at the neurophysiological and molecular levels. Kandel's theory identifies both a molecular mechanism that underlies learning as a function of the contiguity between the CS and the US and — albeit more hypothetically — a mechanism (actually two mechanisms) underlying the *absence* of learning in a situation in which the US occurs without the CS in the intervals between simultaneous (contiguous) presentations of both the US and the CS. I have already presented the outline of a molecular explanation for the creation of an association between the CS and R originally triggered by the US: it is a variant of sensitization that requires a precise sequence in the order and temporal interval between the presentations of the CS and the US. Experimental research with *Aplysia* has shown that, if the CS is presented about half a second before the US, Ca^{2+} channels are opened when the US signal arrives, increasing the signal transmitted by the synapse to the motor neuron (see Kandel 1995, 677). But Hawkins and Kandel (1984) also proposed two molecular mechanisms that reductively explain the phenomenon discovered by Rescorla (1968) and Kamin (1969) described above: learning does not occur when there are isolated occurrences of the US in addition to contiguous presentations of the CS and the US.

Rescorla and Wagner (1972) suggested that this situation is equivalent to *blocking*. According to them, the reinforcement of a complex CS *AX* — composed of simple stimuli *A* and *X* — presented just before the US depends on the total strength that the two components *A* and *X* possessed prior to learning. The phenomenon of blocking consists of the fact that "prior conditioning of *A* reduces the degree to which reinforcement of an *AX* compound increments the associative strength of *X*" (Rescorla and Wagner 1972, 77). In cognitive terms, their theory explains this by the fact (which emerges from their equations) that the variation of the associative strength of the CS *X*, ΔV_X, is proportional to $\lambda - V_{AX} = \lambda - (V_A + V_X)$ (i.e., the difference between the maximum strength λ, which depends on the US used, and the total associative strength V_{AX} of all the stimuli present). Therefore, if V_A is already close to λ because of its previous conditioning, then the presence of *AX* before the US will not significantly increase either *A*'s or *X*'s associative strength. In this case, ΔV_X is close to 0, because V_A is close to λ, whereas V_X is close to 0 because *X* has not been involved in previous conditioning.

Hawkins and Kandel offer a molecular reduction of this blocking phenomenon. During conditioning of *A* (called CS_1 in Hawkins and Kandel 1984), the facilitative interneurons trigger the response more and more exclusively following the presence of *A*, at the expense of its triggering by the presence — immediately afterward — of the US. This is explained by accommodation and recurrent inhibition (Hawkins and Kandel 1984, 385). In the end, the US no longer elicits R, the response being monopolized by *A*. When the complex stimulus *AX* (CS_1 CS_2 in Hawkins and Kandel 1984, 386) appears at the second step of the type of learning that manifests blocking, the presence of stimulus *X* (CS_2) is not followed by the activation of facilitator neurons, which would be necessary for the conditioning of *X*.

Hawkins and Kandel propose to follow Rescorla and Wagner's hypothesis that the situation outlined above, in which isolated presentations of the US alternate with presentations of the CS in contiguity with the US, constitutes a variant of the blocking situation. In the molecular reducing theory, the absence of learning in this situation is explained by the hypothesis that an intermittently presented US produces "conditioning to background stimuli," which "cause continuous excitation of facilitator neurons, rendering them insensitive to the US" (Hawkins and Kandel 1984, 388, 387).

The explanation of the phenomena of blocking and the absence of learning, when there is no reliable correlation between the CS and the US, can therefore be completed by highlighting the underlying neurophysiological processes. This does not mean that cognitive explanations become superfluous. The complex properties of learning by conditioning shown by Rescorla and others can be understood only in cognitive terms, not in purely neurophysiological terms. The identification of the underlying biochemical mechanisms leads one to justify, and sometimes modify, their cognitive explanation. As with any reduction, its fruitfulness is measured by its capacity to induce modifications in the reduced theory as well as in the reducing theory. The discovery of a new phenomenon at the psychological level might require the modification of neuronal and molecular theory, but in the same way aspects of conditioning first discovered at the molecular level might require the modification of psychological theory.

According to some, the discovery of the mechanisms underlying learning by conditioning is part of an evolution that leads, in the long run, to the replacement of psychological theories of learning, and of the psychological concepts involved, by purely neurophysiological theories and concepts.[53] However, the discovery of a reductive explanation of a psychological phenomenon does not lead to its "elimination" as a psychological phenomenon. On the contrary, the discovery of the processes underlying the phenomena of learning strengthens our reasons for believing that these phenomena exist. This is particularly clear when the discovery of the underlying neurophysiological mechanism makes it possible to explain a phenomenon in the precise form attributed to it by the psychological theory subject to reduction. This is illustrated by the mechanism underlying classical conditioning, which explains why the simple contiguity of the CS and the US is neither necessary nor sufficient for conditioning. The elimination of psychological concepts is not justified either when the discovery of the underlying neurophysiological mechanism leads to a correction of the psychological theory: it continues to use the psychological concepts of the CS and the US.

If the reduction of a psychological theory does not lead to the elimination of the psychological phenomenon, then we might be tempted to conclude that it leads to its *identification* with the underlying neurophysiological mechanism (see Causey 1977; Churchland and Churchland 1994). The reduction shows that the cognitive capacity — that of learning to react (with R) to the CS as if it were the US — is identical to a microscopic property, in this case the property of being in a state of sensitization of the pre-synaptic termination of the sensory neuron (originating from the CS), which has synapses with the motor neuron leading to R or with interneurons leading to R. In a similar way, it might be said that the exercise of the capacity is identical to the unfolding of the underlying mechanism.

53 This is the thesis of Bickle (2003). Gold and Stoljar (1999) call it the "radical doctrine of the neuron" (after Barlow 1972).

It is one of the central theses of this book that such an identification is justified neither with regard to the relationship between the temperature of a gas and the kinetic energy of the molecules of which it is composed, nor with regard to the relationship between water and the H_2O molecules of which it is composed, nor finally with regard to the relationship between the conditioning process and the underlying microscopic processes. At first sight, it might seem that the reduction of the (macroscopic) property of being water shows that it is *identical* to the (microscopic) property of being H_2O, in the same way that it might seem that the reduction of the (macroscopic) property of having the temperature T shows that it is identical to the property of being composed of molecules having an average kinetic energy E_{kin}. In the first case, this appearance is the result of an ambiguity in the expression "is H_2O." The property of being a molecule of H_2O is microscopic and can only belong to molecules. Yet the property of *being made up of* H_2O molecules is macroscopic. The reduction of the property of being water shows that this property is identical to the second, which is macroscopic, but not to the first, which is microscopic. Similarly, the reduction of the macroscopic property of having temperature T does not lead to the identification of this property with the microscopic property (of the individual molecules) of having the average kinetic energy E_{kin} or with the property of being a set of molecules whose average molecular energy is E_{kin}: many sets of molecules have an average kinetic energy without having a temperature because they do not interact with each other (see Kistler 1999c). The reduction of the macroscopic property of having temperature T leads to its identification with the macroscopic property of having microscopic components whose interaction allows them to exchange energy and whose average kinetic energy is E_{kin}. The details of the reduction show how temperature is determined by the microscopic properties of the components of the object that has the temperature and by the interactions among these components. In the same way, the reduction of cognitive properties and processes — such as the disposition to learn by conditioning, learning by conditioning itself, and the state of being conditioned in a given way — does not lead to their identification with microscopic properties and processes. Such a neurocognitive reduction shows that the cognitive property of an organism is identical to the property of having parts articulated in a given way so that the neurophysiological properties of the parts and their articulation determine — in a nomological way — the cognitive property of the organism. For example, the reduction of an organism's cognitive state of having learned an association between the CS and the US shows that this is the property of having neurophysiological parts articulated in a certain way. The reductive explanation shows that the neural properties of certain parts of the organism determine the overall property of the organism.

The anti-reductionist conclusion of Gold and Stoljar that "the claim that Kandel's model is a reduction of classical conditioning . . . cannot be sustained" (1999, 825) is based on a conception of reduction as identification. Yet, when we construe reduction as the demonstration of a non-causal relationship of determination[54] of the global properties of a complex system by the properties of its parts and their interactions, it can be argued both that Kandel's model succeeds in reducing[55] certain forms of classical conditioning and that "the concept of synaptic change cannot capture the concept of information or surprise" (Gold and Stoljar 1999, 825). The concepts of information and surprise belong to the cognitive level and apply to the organism, whereas the concept of synaptic change belongs to the neurophysiological level and applies to parts of the organism that play a key role in the reductive explanation of conditioning. The reductive explanation shows in detail how each episode of learning unfolds. Concepts such as information and surprise used in Rescorla's (1988) theory of conditioning can explain, from an evolutionary perspective, why the conditioning process obeys the laws that I have outlined above. To use Dretske's (1988) distinction, reduction allows us to understand the mechanism of conditioning in terms of its "triggering cause," whereas we need to use the notion of information to give a teleological and functional explanation of this mechanism in terms of its "structuring cause." Natural selection helps to explain the appearance of this learning mechanism during evolution: animals capable of conditioning can adapt to their environments because conditioning enables them to act in ways appropriate to the presence of the US even before it is perceived, insofar as the CS objectively contains the information that the US will occur.

Nagel imposes "non-formal" conditions for the success of a reduction.[56] In the ideal case, a reduction induces new hypotheses and research direc-

54 The determination of the properties of a complex system by the properties of its parts and their relationships cannot be causal because it does not extend in time: it is a form of simultaneous determination, whereas causes must precede their effects. I will return to this point in Chapter 4.

55 More precisely, it is a reductive hypothesis that leaves open the question of its truth.

56 "For a reduction to mark a significant intellectual advance, it is not enough that previously established laws of the secondary science be represented within the theory of the primary discipline.

tions, both in the reduced macrotheory and in the reducing microtheory. Many "conservative" reductions — which do not lead to the elimination of the reduced theory — have had the effect of inspiring improvements in both the reduced theory and the reducing theory. However, this situation is conceivable only if each of the two theories is situated within and explains a proper domain of phenomena. The CHB reduction model seems to exclude this possibility: insofar as it is possible to construct an adequate description of macrophenomena within the microtheory, the macrotheory loses the autonomy necessary to inspire new hypotheses or corrections to the hypotheses of the microtheory. Depending on the quality of the analogy between T_R^* and T_R, the old macrotheory is eliminated (if the analogy is bad) or preserved approximately (if it is good). But in the latter case what is retained, strictly speaking, is T_R^*, which has no autonomy in relation to T_B. Insofar as T_R differs from T_R^*, it is false, but false theories cannot inspire corrections to correct theories.

My analysis of the reduction of thermodynamics shows that there are reasons to abandon the requirement that T_R^* be derivable from T_B without linking assumptions. According to the synthetic model of reduction introduced above, the deduction of T_R^* from T_B is not a logical derivation but involves non-analytical laws. If we assume that the macroproperties are determined by the microproperties and their interactions by virtue of laws of nature, and not only by virtue of logical and mathematical rules of calculation, then it seems to be legitimate and necessary to pursue the development of theories at both levels in parallel in order to improve the reduced theory, the reductive theory, and our knowledge of the laws of composition used in reduction. To describe this situation, Robert McCauley introduced the notion of "co-evolution" of theories that deal with the same phenomena but at different levels. He distinguished three variants. Only one of them, "co-evolution$_P$," gives rise to what he called "explanatory pluralism" (1996, 27), in which theories at different levels influence each other. This typically leads to the emergence of a new "interfield theory"[57] that forges a synthesis of the reduced theory, the reducing theory, and the links of determination between them.

The theory must also be fertile in usable suggestions for developing the secondary science, and must yield theorems referring to the latter's subject matter which augment or correct its currently accepted body of laws" (Nagel 1961, 360). Also see McCauley (1981); Enc (1983).

57 This concept was introduced by Maull (1977) and Darden and Maull (1977).

In contrast, the CHB model only allows for the possibility of co-evolution$_M$, in which the reduced theory is justified by its derivation from T_B, but preserves no conceptual independence from T_B, and co-evolution$_S$, in which the "reduced" theory is eliminated. We have seen that, in the cases that we have examined, the conditions for reduction of the co-evolution model$_M$ are too strong. Co-evolution$_S$, which according to Paul and Patricia Churchland is the most appropriate model for reducing psychology to neuroscience, appears from this perspective to be the result of a "category mistake" (McCauley 1996, 34). It seems to be plausible for one theory to eliminate another only when these "theories compete for the same logical space" (Endicott 1998, 59)[58] — that is, seek to account for the same phenomena. Now a microtheory that reduces a macrotheory does not meet this condition, or it could meet it only if the reduction conformed to the CHB model. If the first stage of the reduction consisted of a derivation of T_R* without recourse to linking hypotheses, then the microtheory would cover, through its implication of T_R*, the same domain of phenomena as T_R. However, insofar as the resources of T_B alone are not sufficient to cover the macrophenomena in the domain of T_R and T_R*, T_B and T_R (as well as T_B and T_R*) are not in competition. This removes the plausibility of the idea that T_B can eliminate T_R, even when there is a reduction.

The correction of theories in the course of their unification or reductive integration can be reciprocal (Schaffner 1993, 427–29). It is not always only the reduced theory that is corrected, as suggested by the CHB model: in the case of reductions that have been achieved, the higher-level reduced theory suggests as many avenues of research in the microtheory as the latter suggests in the former.

As we saw above, Rescorla (1968) established at the psychological level that learning by classical conditioning is impossible when, between perceptions in which the CS appears to be associated with the US, the US appears alone, unaccompanied by the CS. This phenomenon, first discovered at the cognitive level, prompted Hawkins and Kandel (1984) to look for a cellular mechanism that could provide a reductive explanation. The fact that the reduced theory still contains elements that can suggest research at the level of the reducing theory, even after the reduction has been completed, undermines the CHB model. According to that model, the reduced theory does

58 This analysis is inspired by Wimsatt's (1976a, 222) comparison between reduction between levels and reduction within a level.

not retain sufficient conceptual autonomy to be able to inspire research at the level of the reducing theory. If the psychology of learning lost all conceptual independence as a result of its reduction, then how could it be a fruitful source of research in neuroscience? Endicott takes this reasoning a step further: insofar as the reducing theory is influenced by constraints "from above" (i.e., from the reduced theory), "the basic reducing theory becomes permeated with high-level concepts and concerns" (1998, 65; see also Gold and Stoljar 1999; van Eck, Looren de Jong, and Schouten 2006).

Functional concepts from molecular biology — such as signal sequence (a sequence of amino acids containing a protein, which has the function of directing the protein to its destination), antibody, secondary messenger, and receptor protein — can be "incorporated in an *integrated interlevel* theory" (Kincaid 1990, 590). However, these concepts cannot be "reduced" to molecular biology (in the sense of being replacable in principle by concepts from the latter[59]) because the explanation of the mechanisms underlying the exercise of these functions requires the use of other macroscopic concepts of cell biology. It is possible in principle, for example, to specify the molecular composition of any antibody. But there are no molecular-level properties common to all antibodies, of which there are millions.[60] The only property that they have in common is the functional property of establishing a bond with an antigen so that this bond triggers an immune reaction. Identifying the underlying mechanism offers no prospect of eliminating the concept of antibody, which alone makes it possible to express a regularity at the macroscopic level, invisible from the point of view of the multitude of underlying microscopic mechanisms.

With regard to research on the mechanisms underlying vision, Bechtel concludes that "there is no basis for assuming that one can provide a complete account of the functioning of the mechanism in terms of the parts alone. The behaviour of the mechanism depends not just on the parts but how they are organized and the context in which they are situated" (2009, 559–60).

59 Kincaid presupposes "the root notion of reduction — that one theory can do all the work of or replace another" (1990, 590).

60 This observation is reminiscent of the one that I made earlier about the failed attempt to reduce thermodynamics to mechanics without appealing to probability but by indicating the initial conditions that characterize systems whose behaviour is in accordance with macroscopic laws. It turns out that these initial conditions can be specified only in terms of macroscopic concepts of thermodynamics: these initial conditions characterize systems that conform to thermodynamics.

To understand the neurophysiological mechanism of vision in terms of the articulation of its component parts, we need to analyze the function of vision as a whole in an animal's interaction with its environment. If we were to try to understand vision merely from the perspective of its neurophysiological mechanism, then we would tend to forget that the function of vision is to inform the cognitive subject about its environment.[61]

8. Conclusion

The reduction between two theories that study the same domain of phenomena at different levels is a major conceptual tool for understanding the process of the unification of science. The rise of cognitive neuroscience is just the latest episode, albeit a particularly spectacular one, in the process of unifying domains of knowledge concerning different scientific theories. The interpretation of this unification, which merges the formerly separate sciences of psychology and neuroscience into a single theory, is of particular importance insofar as it concerns psychology. There is a long tradition of claiming the autonomy and irreducibility of psychology. The prospect of the reduction of psychology gives rise to particularly intense fears and hopes. Given the importance that we attach to our minds, we might fear that such a reduction would reduce us to the level of mere assemblies of cells and thus risk undermining our moral dignity. But we can also hold out hope that we will finally gain understanding and explanation of the mysteries surrounding our minds, such as the origin of mental illnesses, their dependence on certain brain dysfunctions and new ways of curing them, and the function and significance of sleep and dreams. One of the aims of this book is to show that the prospect of reducing psychology to the neurosciences appears to be dramatic and worrying only when viewed under particular interpretations of what a reduction is. Others are compatible with the intuition of the autonomy of psychology and with the existence of a mind or, more precisely, with the existence of cognitive and mental properties distinct from the neurophysiological properties of our brains.

61 The fact that perception depends as much on the neurophysiological mechanism as on the interaction of the cognitive subject with its environment led Clark and Chalmers (1998) to the "extended mind" hypothesis. According to this hypothesis, cognition is extended beyond the cognitive subject to include the environment. See Clark and Chalmers (1998); O'Regan and Noë (2001); Clark (2008).

In the tradition of logical empiricism, the reduction of one domain of phenomena to another is conceived of as an explanatory relationship between the theories that cover these domains of phenomena. The reduction from one theory to another consists of a deductive-nomological explanation: each of the axioms and principles of the reduced theory is deduced from premises taken from the reducing theory. In his now canonical presentation of this conception, Ernest Nagel introduces the distinction between homogeneous and heterogeneous reductions. The reduction of psychology to neuroscience belongs to the category of heterogeneous reductions. In such cases, the reduced theory contains concepts that do not appear in the reducing theory: neurophysiology, for example, knows nothing about motivation, perceptual discrimination, or iconic memory. The debate on the interpretation of heterogeneous reductions hinges on the status of linking statements, or "bridge laws," that must be introduced if we hope to find a deductive explanation of the axioms of the reduced theory, on the basis of the reducing theory. These linking statements are neither metalinguistic and analytical nor, in general, identity statements. In the following chapters, I will examine three other important hypotheses regarding linking statements. According to the hypothesis of conceptual reductionism analyzed in Chapter 2, linking statements can be deduced a priori from the reducing theory alone. According to the hypothesis of classical emergentism analyzed in Chapter 4, linking statements are primitive and inexplicable "transordinal" laws. According to functionalist reductionism, examined in Chapters 3 and 5, linking statements define the conceptual relationship between a functional role and what occupies that role.

I have suggested that linking statements are non-causal laws of a particular type, which I propose to call "composition laws." These are laws that determine the global properties of a system according to the laws that govern the properties of its parts and their interaction. We will return to this concept in Chapters 3 and 4. I have motivated and illustrated it here with two examples: the reduction of temperature to mechanics and the reduction of learning and memory to neurophysiology.

Nagel's model has been mostly criticized for overlooking the fact that historical reductions are only rarely conservative. In general, the reduction of a theory is accompanied by its correction. The main motivation for seeking a reduction is the hope of improving the existing theory. However, insofar as a reducing theory corrects the higher-level theory, as it was before the reduction, it is impossible for the latter to be logically deducible from the reducing theory. We have seen that there are reductions in which the theory T_R* that can be deduced from the reducing theory T_B is not identical but only structurally analogous to the reduced theory T_R. However, I have questioned the thesis of Nagel's critics according to which it is possible to deduce T_R* from T_B without using bridge laws. In particular, I have shown that the reductionist explanation of T_R* from T_B is not an intratheoretical reduction, as predicted by the model put forward by Churchland, Hooker, and Bickle. When we examine historical cases of reduction, it turns out that the reduced theory T_R* is not derived from the assumptions of T_B alone. In the case of the reduction of thermodynamics to classical physics, we have seen that the use of bridge hypotheses is indispensable. Similarly, the reduction of memory fixation and learning by conditioning is intelligible only within the conceptual framework of the reduced theory T_R (or T_R*). The reduced theory is the starting point for the reduction and guides the search for a reducing theory. To take account of the a posteriori nature of the discovery of the laws of composition between the levels of T_B and T_R*, I have suggested a two-part model of reduction. The first or "interlevel" part of this synthetic model of reduction corresponds to the composition laws, discovered on the basis of prior knowledge of T_R and T_B. The second or "intralevel" part corresponds to the demonstration that there is a structural analogy between the theory T_R* deduced from T_B, thanks to the laws of composition, and the reduced theory T_R.

The thesis that the theory T_R* (deduced from T_B and the composition laws) might be only structurally analogous to the reduced theory T_R, without being strictly identical to it, allows us to explain the possibility of reducing multi-realizable properties. In their case, the T_R theory undergoes a separate reduction in each type of system to which T_R applies. The theories $T_R{}^*{}_1$, $T_R{}^*{}_2$, and so on, deduced from different reducing theories T_{B1}, T_{B2}, and so on, and specific to different types of systems, are all analogous to each other and to the reduced theory T_R, without being identical either to each other or to T_R. This is all the more important in the case of psychology: for example, the physiological diversity of the different animal species to which the theory of learning applies provides the main reason for holding the latter to be irreducible to neurophysiology. Given that the structural similarity of the reducing theories — specific to each species — to the reduced theory is sufficient for a reduction, the diversity of neurophysiological substrates is no longer a reason for holding psychology to be irreducible.

The fact that general psychology remains different from — albeit structurally analogous to — species-specific theories also helps to explain why it retains a certain autonomy, even once it has been reduced. This autonomy is essential to explain the fact that discoveries made at the level of the reduced theory often inspire modifications in reducing theories. The observations made at each level with the help of concepts specific to that level are indispensable constraints on the development of interlevel theories. Cognitive neuroscience is such a theory, where items of knowledge obtained separately at the psychological and neurophysiological levels influence and illuminate each other mutually.

2

Can Reductive Explanations Be Constructed A Priori?

1. Introduction

The debate on the nature of the mind and its relation to the body in contemporary analytical philosophy starts from a thesis, the truth of physicalism, and an observation, the existence and relative autonomy of psychology. Physicalism is the ontological doctrine according to which (1) everything that exists either belongs to one of the categories studied by physics or is composed entirely of parts that belong to one of these physical categories, and (2) all of the objective properties of the entities recognized in (1) are either properties studied by physics or reducible to them. Moreover, the very existence of scientific psychology seems to show that there are domains of psychological phenomena within which it is possible to discover regularities independently of the underlying physiological or physical phenomena. This apparent autonomy of psychology seems to suggest that it is irreducible to neuroscience and, even more so, to physics. In Chapter 1, I considered an initial influential argument for the irreducibility of psychology developed within the functionalist conception of mental states: Putnam (1967) and Fodor (1974) argued that psychological properties are multi-realizable, whereas reduction requires bridge principles whose existence is incompatible with multiple realizability.

Davidson (1970) advanced a second argument for the irreducibility of psychology to physics: namely, the conceptual framework of intentional psychology is radically heterogeneous with respect to the conceptual framework of physics and physical sciences, insofar as the attribution of psychological predicates to persons and the attribution of physical predicates to the same persons obey incommensurable criteria of correctness. The application of a physical predicate is governed by an experimental procedure and in the simplest cases by a measurement. The attribution of a psychological predicate must obey a constraint of an entirely different kind: it must make the person to whom the predicate is attributed appear to be rational. Similarly, the criteria for evaluating explanations belonging to these two conceptual frameworks, mental and physical, are radically different. The attribution of mental states, and the explanation of actions in psychological terms, are subject to norms of rationality: for an action to be rational, the means chosen must be adequate given the agent's order of preferences and set of beliefs.[1] In contrast, the standards of correctness for the attribution of physical properties, as well as for physical explanations, are essentially agreement with observation and logical validity, within the deductive-nomological model of explanation.

For these two reasons, it has often been taken for granted that there can be no psychophysical laws, and that psychology is irreducible to physics in principle.[2] Yet this conviction seems to be in contradiction to the doctrine of physicalism, according to which all real properties, in contrast, are reducible in principle to physics. The philosophy of psychology is thus faced with the challenge of finding a way to reconcile the acceptance of physicalism with the autonomy of psychology.

The thesis of the supervenience of psychological properties on physical properties seemed to be able to reconcile physicalism with the irreducibility of the mind. Among the many concepts of supervenience that have been explored, strong supervenience has emerged as the most promising to characterize the relationship between psychological and physical properties. For any set of properties $\mathcal{M}$ and any set of properties $\mathfrak{R}$, $\mathcal{M}$ strongly supervenes on $\mathfrak{R}$ if and only if, necessarily, for any property $M \in \mathcal{M}$, for any x, if x is M, then there exists a property $P \in \mathfrak{R}$, such that x is P and, necessarily, for any y, if y

1 A traditional way of analyzing the rationality of an action is in terms of a practical syllogism. It is rational for agent X to do A if and only if X's doing A is the conclusion of a syllogism whose most important premises are (1) X wants B, where B is a desire of X's that X gives priority to under the circumstances, and (2) X believes that performing A under the circumstances is an adequate way to obtain B. The fact that an action is rational in this sense does not prevent the representations of the reasons for performing it from also being the causes of the bodily movement constituting the action. See Davidson (1963); Kistler (2006c). This thesis is opposed to a traditional doctrine according to which the explanation of an action in terms of its reasons belongs to a conceptual framework incompatible with that of causes.

2 Block speaks of the "anti-reductionist consensus" (1997, 107).

is P, then y is M.[3] One consequence of strong supervenience is that, if the psychological properties of a person supervene on the physical properties of her body (and her environment[4]), then it is impossible for there to be two persons whose bodies (and environments) share all of their physical properties but differ in one of their psychological properties. The concept of supervenience has emerged as a promising tool for reconciling the autonomy of psychology with physicalism, insofar as the relation of supervenience, even strong supervenience, is very weak. In particular, the systematic correlation between the underlying properties $\mathfrak{R}$ and the supervening properties $\mathcal{M}$ is compatible with the absence of psychophysical laws.

However, those who hoped that the use of the concept of supervenience would be sufficient to reconcile physicalism with the irreducibility of the mind have been disappointed.[5] As Horgan (1993) and Kim (1993a) have shown, the strong supervenience of the set of mental properties (and the corresponding states of affairs) on the set of physical properties (and the corresponding states of affairs) imposes no constraint on the origin of their correlation; strong supervenience does not guarantee the truth of physicalism. This is clear from the definition of physicalism given above: the reducibility of all properties to those of physics is part of it. However, supervenience is compatible with dualistic and thus anti-physicalist metaphysical theories, notably with parallelism or occasionalism: if God's intervention guarantees a perfect correlation between physical and mental properties, then the latter supervene on the former. This shows that the postulate of supervenience alone does not require the physical to determine the mental, nor does it require the mental to depend on the physical: in some dualist doctrines compatible with supervenience, all properties are determined by God's will and are dependent only on it.

As long as the nature of $\mathcal{M}$ properties is not specified, the necessary correlation of $\mathcal{M}$ properties with $\mathfrak{R}$ properties is compatible with the radical heterogeneity of $\mathcal{M}$ properties with respect to $\mathfrak{R}$ properties, as in classical dualism, reinterpreted in terms of properties. In other words, the existence of a universal correlation between mental and physical properties — even a

3 In symbols: $\Box\ (\forall M \in \mathcal{M})\ (\forall x)[Mx \rightarrow (\exists\ P \in \mathfrak{R})(Px \wedge \Box\ (\forall y)\ (Py \rightarrow My))]$.

4 If the environment is not mentioned, then the thesis becomes that of local supervenience. We will come back to the distinction between local and global supervenience in Chapter 3.

5 I have developed this point elsewhere (Kistler 2004b) and will return to it in Chapter 4.

necessary correlation such as those in strong supervenience — contains no indication of the origin or explanation of this correlation.

The conception developed in this book overcomes this difficulty by conceiving of the relationship between the physical and the mental on the model of nomological determination, in virtue of non-causal laws of composition. Nomological determination thus appears as the metaphysical foundation of supervenience and allows for its explanation.[6]

Based on the inadequacy of the concept of strong supervenience to express the doctrine of physicalism, a number of authors pursue a completely different strategy to achieve a satisfactory conception of the relation between body and mind: they conceive of the connection between physical and psychological truths as even closer than the necessary correlation of strong supervenience.[7] These authors develop the idea that psychological propositions are merely redescriptions of physical states of affairs in another vocabulary. The psychological conceptual framework allows us to redescribe, with different concepts, the same set of states of affairs that appears as physical when described with physical concepts. The relations of "logical supervenience" (Chalmers), "strict implication" (Kirk), or "entailment" (Jackson) are supposed to ground the physicalist determination of psychological states of affairs by physical states of affairs while avoiding the seemingly mysterious necessity that is part of the concept of strong supervenience.

The general strategy of the proponents of "conceptual reduction," as I propose to call it, is to ground the physicalist determination of the mental by the physical, no longer in a form of natural necessity (compatible with dualism) but in a necessity of conceptual origin. According to Kim (1998), there are no mental properties, only psychological concepts, which are second-order concepts; according to Chalmers and Jackson (2001), psychological concepts are such that one can determine a priori which states of affairs (formulated in physical terms) they apply to, provided that one possesses

6 See Chapters 3 and 4. Broad (1925) attributes to emergence a characteristic often taken — wrongly, as we have just seen — to be an essential component of the concept of supervenience: the dependence of supervenient properties on the properties in their base (or the determination of supervenient properties by their base properties). The definition of supervenience does not, in fact, guarantee such dependence or determination.

7 This strategy forms the common thread of otherwise different conceptions of the mind in nature that have been proposed by Yablo (1992, 1997); Chalmers (1996); Jackson (1998); Kim (1998); Chalmers and Jackson (2001); Kirk (2001); and Esfeld and Sachse (2011).

a complete microphysical description of the actual world. Rather than putting the problem of understanding the relationship of the mental and the physical in terms of different kinds of properties, they conceive of it in terms of the relationship between true propositions (or "truths") expressed in mental vocabulary and true propositions expressed in physical vocabulary. According to this approach, the link between the physical and the psychological is not a natural link but a conceptual one. This implies that it is possible, in principle, to obtain knowledge of any non-physical state of affairs (e.g., a mental one), from a complete knowledge of physical states of affairs, without further empirical investigation (i.e., in a purely a priori manner).

A Laplacian demon[8] that knows the set *P* of all physical states of affairs could extract the set of all other — in particular mental — states of affairs, solely via a priori conceptual analysis regarding *P*. According to Chalmers, this is possible even if *P* contains only microphysical states of affairs: "Laplace's demon, say, who knows the location of every particle in the universe — would be able to straightforwardly 'read off' all the biological facts, once given all the microphysical facts" (1996, 35). According to these authors, the fundamental thesis of physicalism is that the set of all physical states of affairs determines the set of all states of affairs, including, in particular, the set of mental states of affairs. Jackson expresses the thesis by saying that "the psychological account of our world is entailed by the physical account of our world" (1998, 24).[9] To use Kim's metaphor, having created the set of physical states of affairs, God had no further work to do in creating all mental states of affairs.[10]

---

8 A hypothetical being with unlimited reasoning and memory capabilities that allow it to know an exhaustive description of the world at the microphysical level, and to calculate from this description, as well as from the laws of nature, both the future and the past is called a "Laplacian demon":

> An intelligence which, at a given moment, would know all the forces of which nature is animated and the respective situations of the beings which compose it . . . would embrace in the same formula the movements of the largest bodies in the universe and those of the lightest atom; nothing would be uncertain for it, and the future, like the past, would be present to its eyes. (Laplace 1825, 32–33)

Here I am concerned not so much with the power to calculate the future and the past as with the power to derive a description of a state of affairs in macroscopic terms from its description in microscopic terms, at the same instant.

9 Chalmers also speaks of the "logical supervenience" (1996, 33) of the set of all states of affairs on the set of physical states of affairs, whereas Kirk (1996, 2001) says that the latter "strictly implies" the former.

10 This metaphor is often used, for example by Chalmers (1996, 35).

Here is how Jackson defines this "minimal physicalism": (J) "Any world which is a *minimal* physical duplicate of our world is a duplicate *simpliciter* of our world" (1998, 12). By a "minimal physical duplicate of our world," Jackson means a world that is perfectly similar to our own in all physical respects. The qualifier "minimal" means that such a world contains nothing more than what is necessary given its physical constitution.[11]

The purpose of this chapter is to question the possibility of deducing non-physical truths a priori from a description of the world in microphysical terms. The thesis of a priori deducibility from the complete microphysical description *P* is meant to hold for all non-microphysical truths. It can therefore be challenged, without entering into controversies about the specificity of truths about the mind, by focusing on common-sense truths, such as "Water covers most of the Earth" (Jackson 1998, 73).

We will see that the truth of physicalism is not sufficient to guarantee the possibility of deducing macroscopic common-sense truths a priori from *P*, because knowledge of *P* is not sufficient for the construction of their reductive explanation; indeed, such a construction has an a posteriori part that goes beyond knowledge of *P*.

2. A Priori Reduction in the Framework of Two-Dimensional Semantics

Consider this macroscopic fact expressed with common-sense concepts:

(*) Water covers most of the Earth.

According to Chalmers and Jackson, facts of this kind can be inferred a priori from two premises:

(1) a complete description of the state of the world in microphysical terms, and

11 Jackson points out that (J) expresses contingent global supervenience: the truth of the physicalist thesis is contingent insofar as it bears only on the actual world, not on all possible worlds. It is compatible with physicalism that other worlds contain non-physical substances. Kirk proposes another definition of minimal physicalism in terms of the "strict implication" of all states of affairs by the set of physical states of affairs. See Kirk (1996, 246; 2001, 544–45).

(2) an analysis of the concepts used to express the fact in question.

Such an inference produces what Chalmers and Jackson call a reductive explanation. According to them, "there is an a priori entailment from microphysical truths to ordinary macrophysical truths" (2001, 316). This means that it is possible to know a priori that the material conditional $P \supset M$ is true, where *P* denotes "the conjunction of microphysical truths about the world" and *M* a common-sense truth about macroscopic objects and properties, such as water, for example (*), or life: "There are many living things" (317). Their thesis is that a priori conceptual analysis is all that is required to know that $P \supset M$. In Jackson's terms, "physicalism is committed to the in principle a priori deducibility of the psychological from the physical" (1998, 83). In other words, these authors argue that conceptual analysis makes "armchair metaphysics" possible: according to Jackson (1994), conceptual analysis — which can be carried out "in one's armchair" (i.e., without recourse to experience) — is indispensable and fundamental to metaphysics. To use Horgan's (1984) expression, "cosmic hermeneutics" allows all truths to be derived a priori from a (hypothetical) complete description of the world in microphysical terms.

Chalmers and Jackson seek to establish their thesis within the conceptual framework of two-dimensional semantics (Chalmers 1996, 2004; Jackson 1998). We must be content here with a brief presentation of the fundamental concepts that they use in their argument. Primary intension plays a key role. Generally speaking, the extension of a predicate is the set of objects to which it applies; its intension is a function that determines the extension of the predicate in each possible world. Two-dimensional semantics was originally developed in the context of the semantic analysis of statements containing indexical expressions, such as the words *I* and *here* (Stalnaker 1978; Lewis 1980). In the case of such terms, the intension is a function determined by two factors: the context of utterance and the context of evaluation. When I utter the word *I* on a given occasion, the context of utterance determines, together with the lexical meaning of the word (often called, following Kaplan [1989], the "character" of the word), the reference of the word: namely, in the case of *I*, the speaker. It is therefore the speaker who figures in the content of the proposition expressed. Now let us consider the context of utterance as given. The proposition expressed is therefore well determined. We can then ask ourselves about the modal status of this proposition: is it contingent or necessary? The answer depends on the truth value of the proposition in the set of possible worlds. We therefore need to know the extension (or reference) of the terms contained in the proposition in other possible worlds. "I" is a rigid term (Kripke 1972): that is, given a context of utterance, the reference of the term is the same in all possible worlds where the proposition can be evaluated. Other expressions, especially definite descriptions such as "the fastest man over 100m in 2022," are not rigid and denote different individuals in different possible worlds. For an indexical term, the two factors that determine its extension in other possible worlds — the context of utterance and the possible world in which the proposition is evaluated — are therefore independent; this is why we can speak of two "dimensions" of intension.

Here is the definition of the primary intension of a term: it is the function that associates an extension to each context considered as both context of utterance and context of evaluation. This notion is relevant because the speaker is often unaware, at least in part, of the context of utterance. The speaker might be unaware of certain aspects of the context of utterance that determine the content of the indexical terms and thus of the proposition expressed: she might not know where she is when she says *here* or what time it is when she says *now*. However, insofar as she knows the lexical meaning (the character) of the term that she uses, this does not prevent her from knowing the primary intension of the term (and of the proposition expressed) a priori. We can express the primary intension of the word *now* by a series of conditionals: if the word is uttered on Monday at noon (context of utterance), then it denotes, at the same world (context of evaluation), Monday at noon; if the word is uttered on Tuesday at 10 a.m., then it denotes, at the same world, Tuesday at 10 a.m. In each conditional, the antecedent is a world that could, for all the speaker knows, be the one that the speaker is in, its consequent being the reference of the word in that world.

It is crucial for Jackson and Chalmers' argument to assume that the two-dimensional analysis of intension can be applied to other than indexical terms. Kripke (1972) and Putnam (1975a) have suggested that natural kind terms, such as "water," also have an indexical aspect. This suggestion was later developed by Stalnaker (1993) and Haas-Spohn (1995, 1997) as well as by Chalmers and Jackson. According to this hypothesis, terms referring to natural kinds such as "water" that are not overtly indexical nevertheless possess a "hidden indexicality." Insofar as we are partly unaware of the nature of water, the actual world in which we find ourselves acts as the context of utterance: the actual world determines, together with the lexical meaning of the term "water," the reference of each utterance of the term. Let us say that there are three epistemic possibilities of what the content of the term "water" might be: either our world is such that within it water is H_2O, or more precisely that within it water consists of macroscopic samples composed overwhelmingly of H_2O molecules,[12] or that in it water is XYZ or ABC. Like indexical expressions, I can be unaware of which of these worlds I am in yet know the primary intension of the word a priori: if the actual world is such that water is H_2O (context of utterance), then the extension of the term "water" in that world (context of evaluation) is H_2O. Conversely, if the actual world is such that water is XYZ, then the extension of the term "water" in this world (context of evaluation) is XYZ.

The secondary intension is the function that assigns an extension to a term in every possible world (where these worlds are all taken to be counterfactual, except the actual world), where the content of the term is assumed to be determined either by the linguistic meaning alone or by the meaning together with the context of utterance. Kripke (1972) argued that natural kind terms, like proper names, are rigid. This means that their secondary intension is constant: if the reference of the term "water" in the actual world is H_2O, then it has the same reference in all possible worlds. In other words, even when we consider counterfactual worlds in which certain states of affairs concerning water differ from the actual world, we are still talking about the substance that fills the oceans in the actual world.

Jackson and Chalmers' argument proceeds as follows. We have seen that the primary intension of common-sense concepts, such as water, is accessible to us a priori, through conceptual analysis. The primary intension of such a term corresponds to its "character": it is the linguistic meaning, known a priori to all competent speakers. In the case of "water," this meaning can be abbreviated as "the watery stuff we are actually acquainted with" (Jackson 1998, 75). This linguistic meaning determines, together with the context of utterance, in particular the actual world, the content of an utterance of the term. The primary intension of a term consists of a set of application criteria, meaning that it can be expressed by a set of conditionals: each has as its antecedent the description of a world taken to be actual and as its consequent the

12 This precision will be implied henceforth.

extension of the term in that world. Let PH_2O, PXYZ, and PABC be complete descriptions of all the microphysical states of affairs of three (epistemically) possible worlds that differ only in the composition of the aqueous substance. To know the primary intension of "water" is to know the following conditionals: if PH_2O, then water is H_2O; if PXYZ, then water is XYZ; and if PABC, then water is ABC. The a priori knowledge of the primary intension is essentially conditional in that it is a function, which associates to each context (or world) of utterance an extension in the world of evaluation identical to the world of utterance. To know the value of the function (the extension in the world of evaluation), I must know (a posteriori) its argument (the context of utterance). In other words, I must know what the world of utterance is. According to Chalmers and Jackson,

> if a subject possesses the concept "water," then sufficient information about the distribution, behaviour, and appearance of clusters of H_2O molecules enables the subject to know that water is H_2O, to know where water is and is not, and so on. This conditional knowledge requires only possession of the concept and rational reflection, and so requires no further a posteriori knowledge. . . . Possession of a concept such as . . . "water" bestows a *conditional ability* to identify the concept's extension under a hypothetical epistemic possibility. . . . Because all the relevant empirical information is present in the antecedent of the conditional, empirical information plays no essential role in justifying belief in the conditional. So . . . [this conditional] is a priori. (2001, 323–25)

The primary intension of a concept does not give us its extension, in a given world, but it does tell us how the context (i.e., the nature of a given world) determines this extension. The extension of the term "water" depends on the world of utterance, but knowledge of a physical description of the world of utterance (PH_2O or PXYZ etc.) puts the possessor of the concept "water" in a position to determine a priori the extension of the concept in that world.

As Chalmers and Jackson put it, "if a subject possesses a concept and has unimpaired rational processes, then sufficient empirical information about the actual world puts a subject in a position to identify the concept's extension. . . [A] 'water'-free description of the world can enable one to identify the referent of 'water'" (2001, 323).

Chalmers and Jackson seek to show that *P* (the full description of the real world in microphysical terms) allows us to infer a priori that

(*) Water covers most of the Earth.

The structure of this a priori deduction is as follows. *P* is supposed to contain the information that

(1) H_2O covers most of the Earth.

Then the conceptual analysis of the word *water* yields (this is an analytical and a priori truth) that

(2) water is the watery stuff that we are acquainted with.

Finally, the context of the utterance of (*) — that is, the world in which (*) is uttered — provides us with the information that

(3) H_2O is the watery stuff that we are acquainted with.[13]

(1), (2), and (3) together allow us to derive (*).

The possibility of such an a priori derivation of all macroscopic, common-sense, and scientific truths from a complete description of all microphysical states of affairs, through conceptual analysis alone, has been challenged on various grounds, notably by Block and Stalnaker (1999) and Byrne (1999).

First, (1) contains the macroscopic concept of the Earth. It is therefore necessary to justify the idea that one can derive (1) a priori from *P*, exclusively composed of truths in microphysical terms. This seems to be doubtful for reasons that were presented in Chapter 1 and to which we will return: the

13 This corresponds to the context of the actual world. In another possible world, the context would have determined, for example, this information: "XYZ is the aqueous substance that we are familiar with."

concepts that one uses to describe microscopic objects do not contain information about the macroscopic properties of the objects composed by these microscopic objects.[14] For this reason, the deduction of macroscopic properties from information about microscopic properties alone cannot be a priori.

Second, it is questionable whether the set of all microphysical truths, expressed in the language of "ideally completed physics," is well determined.[15] The concept of a completed or ideal physics is often used, for example, to define the concept of the law of nature.[16] However, the existence of scientific revolutions prevents us from conceiving of "completed physics" as a conservative extension of current physics. There is no reason to think that the concept of completed physics determines a single system of concepts and propositions, rather than a multitude of theoretical systems, all empirically adequate but incompatible with each other. Now, without a well-determined antecedent *P*, the implication $P \supset M$ has no well-determined meaning either, and the question of its a priori character cannot even be asked.

Third, in the remainder of this chapter, I will point out another major weakness of Chalmers and Jackson's argument. The epistemic status of (3) is problematic: Block and Stalnaker have pointed out that "the claim that H_2O is the (or even a) satisfier of the primary intension of 'water' is not a microphysical claim" (1999, 45). Proposition (3) is not part of *P*, so it cannot be used in the premises of a priori deduction in the same way as (1). Nor is it an a priori truth, so it cannot be used in the same way as (2). Block and Stalnaker offer no analysis of the nature and epistemic status of (3). It is important to fill this gap because the success of Chalmers and Jackson's cosmic hermeneutics project depends crucially on the status of (3). If it is true, as I will try to show, that (3) cannot play the role that Chalmers and Jackson ascribe to it, then we have no reason to think that macroscopic truths can systematically be deduced a priori from *P*. Specifically, I question the thesis that *P* conceptually entails (3) or, in Jackson's terms, that "a rich enough story about the H_2O way things are does conceptually entail the water way things are" (1998, 149).

14 See section 4 of this chapter.

15 This objection has been raised by Byrne (1999). See Chalmers and Jackson (2001, 334).

16 This is particularly the case with the so-called best system view advocated by David Lewis (1973, 73). See Kistler (1999b, Chapter 6; 2006d, Chapter 6).

3. Two Concepts of Reduction and Realization: Micro-Macro and Role-Occupant

In order to produce reductive explanations of macroscopic phenomena (with the exception of qualitative aspects of subjective experience, which Chalmers takes to be irreducible), Chalmers argues that it is sufficient to have (1) detailed knowledge of microphysical states of affairs and to have accomplished (2) the "functional analysis" (1996, 43) of macroscopic concepts. The former is empirical in origin, but the latter can be accomplished in a purely a priori manner.

Once the functional analysis[17] of the concept that describes a macroscopic phenomenon has been completed, all that remains to be done is to discover "how those functions are performed. . . . Once the relevant details are in, a story about low-level physical causation will explain how the relevant functions are performed, and will therefore explain the phenomenon in question" (Chalmers 1996, 44).

Chalmers uses the reductive explanation of heat as an example. Heat itself is a physical concept, but according to Chalmers its microreduction follows the same pattern as the microreduction of non-physical macroscopic phenomena, in particular psychological ones. According to the functional analysis of the macroscopic concept of heat, it "is the kind of thing that expands metals, is caused by fire, leads to a particular sort of sensation, and the like" (Chalmers 1996, 44–45). This analysis shows that heat — what was only implicit before the analysis — is "a causal-role concept," which characterizes itself "in terms of what it is typically caused by and of what it typically causes, under appropriate conditions" (Chalmers 1996, 45).

In general, the functional analysis of a concept shows that the concept describes a causal role. Accordingly, to complete the reductive explanation, it is sufficient to discover, in a second empirical step, what fulfills the role thus defined: it is discovered "that heat is realized by the motion of molecules" because "the motion of molecules is what plays the relevant causal role in the actual world" (Chalmers 1996, 45). However, as we will now see, the a priori deduction of (*) from *P* and the functional analysis of the concept of

17 It is analogous to what Kim (1998) calls the "functionalization" of macroscopic concepts. However, Kim specifies that it is the first step in functional reduction, whereas Chalmers says that a reductive explanation is "accompanied" in general (1996, 43) by such a functional analysis.

water is fallacious because it exploits an equivocation about the meaning of the concept of reduction, combined with an equivocation about the concept of realization. Once the ambiguity is removed and the two meanings of "reduction" and "realization" are distinguished, we will see what the conceptual analysis can and cannot achieve. It will also explain why the possibility of "cosmic hermeneutics" appears to be plausible at first sight.

Let us consider the case of heat. According to the analysis of this concept by Chalmers, heat is that which causes certain events and processes (e.g., raising temperature) and that which is caused by certain events and processes (e.g., combustion). The reductive explanation is then accomplished by finding out what fulfills this role (i.e., what "realizes" heat). According to Chalmers, it is thus possible to bridge the distance between a role concept (i.e., a second-order concept) and the first-order concept of what fulfills the role, alongside the distance between a macroscopic property concept and an underlying microscopic property concept such as molecular motion, in a single step.

However, there are in fact two steps to be taken.[18] The functional description defines a role in terms of interactions between macroscopic objects, a role that can be played only by a macroscopic property. The distinction between the role and the occupant of the role is independent of the distinction between the microscopic and the macroscopic: there are macroscopic roles fulfilled by macroscopic properties and microscopic roles fulfilled by microscopic properties. Two theoretical roles contribute to determining the identity of heat.[19] (1) The heat δQ lost by a closed physical system is equivalent to the work δW that it provides,[20] and (2) in a reversible process, the change δQ_{rev} in the amount of heat contained in a system is proportional to the change in its entropy (dS) and to its temperature. (In symbols, $\delta Q_{rev} = TdS$). However, only a macroscopic property (i.e., a property of macroscopic objects) can play these roles.

18 In response to Byrne (1999), Chalmers and Jackson (2001, 334n16) acknowledge that such a deduction must involve two steps. However, their argument for a priori deducibility does not take into account the step corresponding to the reduction from the macroscopic to the microscopic, the discovery of which, as I will show later, is always a posteriori.

19 The word heat designates the property that occupies the role, not the role itself, but it does so by way of a definite description of the role: heat is the property that has such and such functional and causal relations with such and such other properties.

20 Given that the total internal energy U is constant in an isolated system, $dU = 0$ and therefore $\delta Q = - \delta W$.

This point is worth looking at a little closer. The distinction between "macroscopic" and "microscopic" can be taken in a narrow or broad sense. In the narrow sense, an object is called "microscopic" in comparison to a given macroscopic object if it is smaller by several orders of magnitude than the latter. In the broad sense, each constituent part of an object can be called "microscopic" relative to the object as a whole and the object itself "macroscopic" in relation to that part. Heat is an essentially macroscopic property in the sense that it cannot belong to microscopic objects (in the narrow sense): a single atom cannot be hot. It is part of the conditions for the possibility of attributing the property of being hot to an object that the object has microscopic components, preventing it from being attributed to the individual microscopic components themselves.[21]

The first step in the microreduction of heat is to associate a categorical property with the role of heat: a macroscopic property designated by a first-order predicate is discovered, which plays the role, itself designated by a second-order predicate, typically in functional or dispositional terms. The discovery of the microproperties of the parts of the hot object that give rise to the macroproperty that plays the role occurs in a second and independent step: one step can be accomplished without the other.

Let us call them, respectively, RO reduction (RO for role-occupant) and MM reduction (MM for micro-macro). A reduction of the first kind, an RO reduction, leads to the discovery that a categorical property plays a previously determined role. For example, the concept of heat is primarily a role concept; this role can be made explicit by conceptual analysis. The development of thermodynamic theory led to the construction of the concept of heat as a form of energy equivalent to work: this concept was central to Carnot's (1824) theory of the heat engine. The RO reduction identifies internal energy, a categorical concept, as what fulfills the role of heat. The RO reduction is a conceptual reduction[22] that does not involve different properties; it consists

21 "Microscopic" components in the broad sense might themselves have components. What is crucial here is that heat cannot be ascribed to microscopic components in the narrow sense, such as isolated atoms or molecules.

22 It cannot always be accomplished a priori: this is possible only if one already knows the functional description and a categorical description of the property. Therefore, even RO reductions are not a priori in the sense that the reduction can be constructed by using only the categorical basis alone. The functional description of a property cannot by deduced a priori from any of its categorical descriptions.

of discovering that a property known by a categorical description plays a role characterized functionally or dispositionally.

In contrast, an MM reduction, typically the result of a later stage of scientific research on a natural property, brings different properties into relation: properties of a macroscopic object and properties of its microscopic parts. In the case of heat, Boltzmann and others discovered that laws involving heat could be derived from molecular models. This MM reduction of heat was discovered later than its RO reduction, in the 1870s. When I wrote earlier about the reduction of the property of being water and the property of having temperature *T* to the properties of the components of the objects having these properties and their interactions, I was talking about MM reductions.

Take the case of water: it has the functional or dispositional property of being transparent to light.[23] If water in its liquid state is exposed to light, then light passes through it, so that we can see through it. The reductive explanation of this property of water goes through two steps. First, the macroscopic dispositional property of transparency is RO reduced to the macroscopic property of having a certain absorption spectrum of electromagnetic radiation.[24] This property manifests itself in the form of transparency: water is transparent to the rays that it does not absorb.[25]

Second, the absorption of infrared in water is explained (in the form of an MM reduction) by the absorption of "parts" of light by "parts" of water. Individual photons are absorbed by individual molecules provided that their energy (and wavelength) correspond to the characteristic energy of one of the intramolecular vibrations accessible to the molecule given its geometry.[26]

23 Needham (2000) shows that certain macroscopic characteristics are part of the identity conditions of water.

24 Water absorbs rays whose wavelength falls in the centimetric range, then in the infrared (wavelength between 2 and 6 mm), then in the far ultraviolet (wavelength 1,650 Å). See Caro (1995, 86).

25 The property of having a certain absorption spectrum can be conceived as dispositional or as categorical. In Chapter 3, we will see that the distinction between dispositional and categorical is a semantic distinction concerning the meaning of predicates rather than a distinction between types of properties.

26 Symmetrical vibration of the two O atomic nuclei with respect to the H nucleus, anti-symmetrical vibration (where the directions of movement of the O nuclei are opposite), or torsion; absorption in the centimetre wave range is explained by the absorption by the molecules of the energy required for rotations; absorption in the ultraviolet range is explained by the absorption — by the molecular electrons — of the energy required to pass into a molecular orbit that corresponds to an "excited" state of the electron.

When it is said that these microscopic mechanisms "realize" the transparency of water, the word *realize* can have two meanings, which might contribute to obscuring the difference between the two stages of reduction. It is possible to speak of "realization" to designate the two relations: one can say that part of the internal energy of a gas (δQ_{rev}) "realizes" the cause of the increase in entropy, as a function of temperature, according to the formula $\delta Q_{rev} = TdS$. In this context, the word *realization* refers to RO realization, a relation between what occupies the role and the role itself.

But there is another meaning of "realization" that expresses what might be called "micro-macro realization" or "MM realization": it is the relationship between the microscopic properties of the components of an object and a macroscopic property of that object to which the interaction between the components gives rise. It is in the sense of MM realization that Chalmers can say that the motion of molecules "realizes" heat: the motion of molecules is the microscopic property that MM realizes heat as a role-occupant (i.e., as a first-order macroscopic property).

The problem is that only RO realization can be discovered a priori, whereas MM realization is always discovered a posteriori. When both the role and the occupant are known, it can be discovered a priori that they are in a role-occupant relationship. In contrast, the discovery of a microreduction (i.e., the discovery of the microscopic properties and interaction laws that determine the macroscopic property together), is always a posteriori.

The choice to call both of these relationships "realization" can be misleading. In reality, the only thing that they have in common is that each corresponds to one of the two reduction relations that I have distinguished. However, the differences are important: RO realization corresponds to a relation between concepts, just like RO reduction, whereas MM realization corresponds — like MM reduction — to a relationship between distinct properties, microscopic in one case and macroscopic in the other. An MM reduction describes how the microscopic properties of the parts that make up an object naturally determine its macroscopic properties, whereas an RO reduction consists of the discovery of a categorical description of a property first conceived of in a functional way.

4. Multi-Realizability

The concept of realization, like that of implementation, allows us to conceive of the possibility that a property can be realized in different ways: such a property is "multi-realizable." The thesis that mental properties are multi-realizable was introduced in the philosophy of mind in the context of machine functionalism and the analogy of the mind with computer software. Just as software can be "implemented" in different ways in different machines, so too cognitive states can be implemented in brains with different neurophysiological properties.

The analysis of multi-realizability has enhanced the confusion between the two kinds of realization: macroscopic roles in general can be "RO realized" by different categorical macroscopic properties, so that these roles are RO multi-realizable. But macroscopic properties are also often MM multi-realizable too, in the sense that objects can share macroproperties determined, in different cases, by different microscopic properties of their parts.

RO realization allows for the possibility that a single causal role can be occupied by different occupants.[27] Many biological functions are RO multi-realized in the sense that they are performed by different categorical properties in different biological species. The function of enabling an organism to see (i.e., to give it access to the information contained in the light waves that reach the surface of its body) can be fulfilled by several properties. The property of being a mammalian eye and the property of being an arthropod compound eye are two first-order structural properties that perform the function of enabling an organism to see. Either can play the role of giving the organism access to the information contained in light. Being an antibody is a functionally designed (microscopic) property that can be achieved by "millions of different chemical structures" (Kincaid 1990, 585), which are also microscopic. This example illustrates the above-mentioned fact that an RO reduction can be achieved only after the independent discovery of both a functional description (the determination of the role that antibodies play)

27 The reverse is also true. Morange (1998) mentions numerous examples of biological molecules, and in particular genes, that play several roles assumed to have been acquired successively and independently of each other. Tompa, Szasz, and Buday (2005) and Tobin (2010) analyze the case of so-called moonlighting proteins that fulfill several functions, in an analogy to people having a second job at night, in addition to their main job.

and a categorical description (the RO reduction consists of the discovery that a property that satisfies the latter also satisfies the former).

The natural determination of the properties of a macroscopic object by the properties of its parts gives rise to multi-realizability in a very different sense from that of RO multi-realizability: consider, for example, the property of being a hemoglobin molecule. Its overall structure, or "conformation," allows the molecule to play its biological role of transporting oxygen. This way of speaking tends to obscure the fact that there are many different types of hemoglobin molecules in different biological species that differ in their parts: that is, in the amino acids that make up the sequence of each of the four proteins that make up the four subunits of the molecule (a "tetramer"). The amino acid sequence is known as the "primary structure" of the hemoglobin molecule. Each type of hemoglobin has its own primary structure and differs from the other types in some of its 140 amino acids. Only 9 of 140 positions are occupied by the same amino acids in all hemoglobin species. The chemical properties of these nine amino acids and their interactions determine the "conformation" of the molecule.[28] It is this overall structure of the molecule that allows hemoglobin to play its role in all biological species. As Rosenberg states, "it is the quaternary structure that produces haemoglobin's remarkable functions" (1985, 77).

However, this overall structure can be determined naturally by a large number of different properties at the level of the parts (i.e., by all of the primary structures that have in common the nine amino acids at the "strategic" positions). The existence of a single overall structure common to all of these molecules justifies speaking of the (kind of) hemoglobin molecule in the singular. However, taking into account the different microscopic structures also justifies speaking of hemoglobins in the plural (Rosenberg 1985, 77). Each of these microstructures naturally determines the same overall structure. The natural determination thus establishes a "many-to-one" relationship between the microstructures and the overall structure. Using the term "macroscopic"

28 This determination goes through two intermediate steps: the chemical properties of these nine amino acids and their interactions determine where the chain folds or overlaps, giving rise to the "secondary structure," which has the effect of bringing together distant amino acids given their positions in the chain, giving rise to new interactions that determine the "tertiary structure." Finally, the "quaternary structure," which characterizes the overall structure of the molecule, is determined by the interactions between the four subunits that come together in a stable conformation. I will consider this example again in Chapter 5.

in a broad sense, so that it generally characterizes the properties of a whole in relation to the properties of its parts, it can be said, at the risk of confusion with RO multi-realizability, that the macroscopic property of being a hemoglobin molecule can be MM realized by many different microstructures.

The function of transporting oxygen in an organism is not only MM multi-realized but also RO multi-realized (Kurtz 1999). Indeed, in some marine invertebrates, such as brachiopods, hemerythrins perform the oxygen transport function, and in some arthropods and molluscs that role is played by hemocyanins, in which oxygen is bound to a pair of copper atoms rather than being bound, as in hemoglobin, to a heme group around an iron atom (Kurtz 1992; van Holde and Miller 1995; van Holde, Miller, and Decker 2001).

The crucial point for my argument — which might be blurred by the confusion between RO realization and MM realization — is that the discovery of MM realization (i.e., the natural determination of the properties of a whole by the properties of its parts) is always a posteriori. Even if we had an absolutely complete description of a situation in microscopic terms, together with the complete set of microscopic laws that apply to it, we would still not know the MM reduction of the macroscopic properties exemplified by the objects composed from the microscopic objects that appear in the microscopic description.

The reason is that not all of the laws necessary to deduce the macroscopic properties belong to the level of the reducing theory, nor can they be deduced from it. In the important case of the reduction of thermodynamic — and thus macroscopic — concepts, such as heat, entropy, or temperature, their MM reduction to mechanical concepts that apply to the microphysical components of the systems to which the thermodynamic concepts apply depends on the introduction of the concept of an "ensemble" of systems that has no meaning at the microscopic level. The construction of macroscopic concepts such as temperature requires the use of new conceptual tools that have no microphysical equivalent and cannot be constructed with the concepts appropriate for describing microphysical objects and states of affairs.[29]

---

29 See Sklar (1993) and Chapter 1. Chalmers and Jackson are hesitant about the need to include laws, in addition to particular microphysical facts, in the reduction basis. According to Chalmers, "high-level facts are entailed by all the microphysical facts (perhaps along with microphysical laws)" (1996, 71). Surely, no reduction of a macroscopic phenomenon can be accomplished without using the laws that apply to the microscopic entities mentioned in the reducing theory: without the laws governing the interactions between the molecules of a gas, it is

The same is true of the analysis of other cases of successful MM reduction: the reduction of classical genetics to molecular biology provides genetic explanations of macroscopic phenomena, but molecular biology does not claim to construct concepts applying to macroscopic phenomena.[30] In the reductive explanation of certain elementary forms of learning — such as habituation, sensitization, and classical conditioning — no effort is made to eliminate essentially macroscopic concepts such as stimulus, reflex, conditioning, and withdrawal behaviour in favour of microscopic concepts. Even after the reductive explanation of these phenomena, psychological concepts continue to provide the framework within which they are described (see Chapter 1).

The advocate of the thesis of the a priori deducibility of macroscopic truths from a complete (hypothetical) microphysical description *P* of the world can make two rejoinders. A first rejoinder is to incorporate all of the laws necessary for reduction into the set of premises *P*, including those that are not purely microscopic. However, this would trivialize the thesis of a priori implication (i.e., deprive it of its content), insofar as the content of the premises *P* would no longer be exclusively microscopic.[31]

A second rejoinder is to argue that the fact that one can reduce, for example, thermodynamic laws only by making use of irreducibly macroscopic concepts, such as the concept of an ensemble, only shows the inability of *current* physics to accomplish this reduction from a purely microscopic basis, whereas what is at stake is the *possibility in principle* of such a reduction. Here one admits that the actual reductions accomplished in the history of science *start* from the knowledge of macroscopic properties and that their

impossible to deduce the properties of the gas. However, though the microscopic laws are necessary, they are not sufficient for deducing macroscopic facts: only our prior knowledge of macroscopic phenomena guides us in the construction of concepts adequate to their description.

30 See Kitcher (1984), Schaffner (1993), and Morange (1998) for many illustrations of this fact, in the context of determining the macroscopic properties of organisms from the properties of their genes.

31 Chalmers and Jackson allude to the problem raised here, that no MM reduction can be accomplished a priori, when they point out that "the only worry" about the truth of their thesis that describing the world in microphysical terms implies all descriptions of the world in macroscopic terms "might concern the status of bridging principles within physical vocabulary" (2001, 331). Rather than confronting this difficulty, they propose to "bypass" it "by stipulating that the relevant physical principles are built into P" (331). This indeed solves the problem but at the cost of abandoning the thesis initially defended and criticized here, according to which a (purely) microphysical description a priori implies all macroscopic truths.

elaboration depends on the prior knowledge of the macroscopic properties to be explained.[32] However, while admitting this about the *actual* discovery of MM reductions, the defender of a priori reducibility could simply postulate that it is still *possible in principle* to deduce all macroscopic truths from a purely microscopic basis. A future molecular biology will shed the conceptual framework bequeathed to it by classical genetics to become a purely microscopic science, and a future neuroscience will construct reductions that make no use of macroscopic cognitive concepts.

Indeed, no logical inconsistency seems to prevent such a possibility. Meanwhile, the burden of proof lies with those who assert a possibility that does not correspond to actual scientific discoveries of MM reductions. As long as historical reductions do not confirm the existence of inferences of macroscopic states of affairs from purely microscopic premises, it seems to be gratuitous to assert that such a feat is nevertheless possible in principle.

5. Conclusion

The deduction of macroscopic common-sense truths from *P*, the hypothetical complete description of all microscopic states of affairs, necessarily goes through two steps: the first is the discovery of an RO reduction: that is, the discovery of the property that fulfills (or properties that fulfill) a certain functional role. Conceptual analysis allows us to discover which properties play the roles corresponding to common-sense concepts such as "water" or "heat" or scientific concepts such as "oxygen carrier." However, the RO reduction is not, for all that, an a priori deduction from the microscopic description *P*, insofar as both of these properties and the functions that they perform belong to the macroscopic level.

The explanation of macroscopic phenomena in microscopic terms is the subject of a second reduction step, which I have called MM reduction. Historically, and as we have seen in Chapter 1, the premises of MM reductions were not purely microscopic. First, some laws, in particular statistical laws relating macroscopic properties to the properties of microscopic constituents of matter, are irreducible to the laws governing microscopic properties and their interactions. The historical cases of reductions of biological or cognitive phenomena also involved macroscopic concepts not built upon

32 "Building a model . . . is not a matter of deduction" (Holland 1998, 9).

a microscopic basis. Second, these historical reductions were accomplished only through prior knowledge of the macroscopic phenomena to be reduced: they proceeded by constructing a model of the microscopic phenomena under the constraint of its adequacy to the macroscopic phenomena, known beforehand.

The deduction of a macroscopic truth expressed with common-sense concepts from *P*, as envisaged by Chalmers and Jackson, must have two parts, one corresponding to an RO reduction and the other to an MM reduction. Since neither the historical RO reductions nor the historical MM reductions took the form of deductions from the mere knowledge of microscopic phenomena, the burden of proof is on those who proclaim a principled possibility that does not correspond to the reality of historical reductions.

Chalmers and Jackson's argument that macroscopic common-sense truths, such as (*), can be deduced a priori from a complete description of the world in microscopic terms is fallacious because it relies on an equivocation: the concept of reduction is sometimes understood in the sense of RO reduction and sometimes in the sense of MM reduction. Contrary to what Chalmers and Jackson claim,

(3) H_2O is the aqueous substance that we are familiar with

cannot be deduced from *P* a priori, just with the help of conceptual analysis. One of the steps in the reductive explanation of (3) is a local and a posteriori MM reduction that allows us to deduce, from a microphysical description and various micro- and macrophysical laws, the macroscopic properties of the substance composed of H_2O molecules: the facts that this substance is liquid at ambient temperature and pressure (near the surface of the Earth, in summer, not too close to the poles), has a reduced viscosity, is transparent to light, et cetera.

Proposition (3) has a hybrid character, partly microscopic ("H_2O"), partly macroscopic ("the watery substance"). For Chalmers and Jackson's argument to be valid, it would have to be purely microscopic, on the one hand: then it would be plausible that it is a priori derivable from a complete microphysical description of the world. On the other hand, it would have to be purely macroscopic: to be the object of the purely a priori discovery that some macroscopic property plays a certain macroscopic role, both the conception of the role and the conception of its occupant would have to be macroscopic, because the natural determination of the macroscopic by the microscopic cannot be the object of a priori discovery.

3

Cognitive Abilities as Macroscopic Dispositional Properties

1. Introduction

A number of cognitive abilities are the subject of reductive explanations. However, the interpretation of the significance of these explanations remains controversial. In Chapter 1, I analyzed a case of reduction of an elementary psychological ability.

We have seen that neuroscience has discovered the neurophysiological basis of certain forms of simple learning, in particular sensitization and classical conditioning, during which an animal acquires the ability to react in an appropriate and differentiated way to external stimulation. It seems to be natural to consider the result of such learning as the acquisition of a disposition.

The dispositional conception is as appropriate for the mental states of naive or "folk" psychology as it is for the cognitive states postulated by scientific psychology. The meaning of the statement "Mary is intelligent" seems to be equivalent to a series of conditionals about the conditions under which her intelligence manifests itself: if Mary were given a complex mathematical problem, then she would solve it easily. If faced with a difficult choice, then she would act on the basis of a balanced appreciation of the advantages and disadvantages of the various alternatives open to her. Similarly, as we saw in the example of the classically conditioned state of *Aplysia*, the states postulated by cognitive psychology are often characterized dispositionally. For example, "the *Aplysia* has learned to react with CS to the perception of R" means that, "if the *Aplysia* were exposed to CS, then it would react with R."

However, the reality of dispositional properties has been challenged. It is controversial whether vases have a real and causally efficacious property that corresponds to their fragility. In the same way and for the same reasons, it is controversial whether Mary has real properties that correspond to her intelligence or her memories.

A vase falls from the top of the shelf onto the tiled floor. Unsurprisingly, it breaks. Is the fragility of the vase one of the properties that contributes causally to the fact that it breaks? If the answer is yes, then the fact that a mental property is dispositional no longer counts as a reason to deny that it has causal powers. What makes fragility a dispositional property? The concept of a dispositional property is that of a property characterized essentially by a subjunctive conditional or counterfactual.[1] An object is fragile to the extent that, under otherwise normal circumstances, it would break if it fell from some height onto a hard surface. It is not necessary for the event described by the antecedent of this conditional to ever occur: it is possible to have a dispositional property that does not manifest itself, and it is even possible to have one that *never* manifests itself. Dispositional properties are distinguished from categorical properties by the fact that the latter are not characterized with the help of subjunctive conditionals. The property of an object of being spherical is attributed according to a criterion that involves only the actual world: all of the points on the surface of the object are at the same distance from its centre.

Armstrong (1968, 88; 1997, 70–71) defended the idea that dispositional properties can be causally efficacious, in particular by bringing about their manifestation, in a situation in which they are put to the test. In the case of the vase, its falling is a situation in which its fragility is put to the test, or "tested,"[2] and its manifestation is the vase breaking. Prior, Pargetter, and Jackson (1982) sought to show, on the contrary, that dispositional properties (or dispositions[3]) are incapable in principle of causal efficacy. This can be ex-

---

1 The relevant concept is that of subjunctive conditional because it is not necessary for the antecedent to be false in the actual world, which, strictly speaking, characterizes counterfactuals (Mellor 1974). Nevertheless, I will follow the current usage in this debate and speak interchangeably of subjunctive conditionals and counterfactuals.

2 In an otherwise normal test situation (I will come back to this condition later), the disposition manifests itself. See Carnap (1936–37); Goodman (1983).

3 Many dispositional properties have *different* characteristic manifestations in *different* types of test situations. Mellor mentions mass as an example of such "*multi-track*" dispositional properties (2000, 760): in a situation in which its possessor is subjected to a force f, mass m gives it both the disposition to acquire an acceleration f/m and the disposition to exert the force mm'/r^2 on another mass m' situated at a distance r from it. Fragility is another dispositional property that gives

pressed by saying that they are *epiphenomenal*. However, this disagreement conceals a partial consensus on the interpretation of the nature of dispositions: the dispositions at issue in this debate are macroscopic and in principle reducible to a microscopic basis. The terms "macroscopic" and "microscopic" are used here in a broad sense: the properties of a whole are macroscopic in relation to the microscopic properties of its parts. In the controversy over the efficacy of dispositions, everyone agrees that the only causally efficacious properties are microscopic properties of the reduction base. Since it is generally presupposed that these properties themselves are not dispositional, the reduction base is also called the "categorical basis" of the disposition. The disagreement concerns only the question of whether a given macroscopic dispositional property inherits this efficacy because it is identical to its reduction base, which is Armstrong's position,[4] or whether it is epiphenomenal because

its possessor different dispositions: the disposition to break in a certain type of situation and the disposition to crack in another type of situation. Similarly, the high temperature of a flammable gas is a dispositional property that gives the gas not only the disposition to be at a certain pressure but also the disposition to explode. In the case of such properties, the distinction between *dispositional property*, or "power," and *disposition* is important. I will come back to this distinction later in the chapter. Nevertheless, I will often refer to dispositional properties as "dispositions" in contexts in which this is not likely to introduce confusion between the dispositional property and the different dispositions to manifest itself that it gives its possessors.

4 In the general context of analyzing the logic of the reduction of one scientific theory to another, many authors follow Causey's (1977) thesis that the reduced property and the reducing property are *identical*. In Locke's writings, one can find both passages defending the doctrine of the *identity* of dispositions (or, in his terminology, "powers") with their microscopic and categorical basis and passages expressing the idea that dispositions are only *dependent on* this basis, the thesis that I defend in this book. Locke seems to express the first doctrine when he says that "whiteness or redness are not in it [i.e., in porphyry] at any time, but such a texture, that hath the power to produce such a sensation in us" (1689, Book II, Chapter VIII, para. 19). Colours are identical to a texture, a categorical microscopic property of the basis of reduction; properties of the latter category can be causally efficacious when they produce in us the experiences by which colours manifest themselves when we look at coloured objects. But other passages seem to be compatible with the second doctrine: "Colours and Smells . . . and *other the like sensible Qualities* . . . are in truth nothing in the Objects themselves, but Powers to produce various sensations in us, and *depend on those primary Qualities*, viz. Bulk, Figure, Texture, and Motion of parts" (Locke 1689, Book II, Chapter VIII, para. 14); here Locke says that sensible qualities *are dependent on* texture and other "primary qualities": that is, on categorical microscopic properties rather than identical with them. I express essentially the same idea by saying that microscopic properties *determine* macroscopic dispositional properties. To say, as Locke does, that secondary qualities are "powers to produce various sensations in us," seems to be compatible with the thesis defended in this chapter, according to which macroscopic properties can be conceived as dispositions that can cause certain effects, such as sensations.

it is distinct from it, the so-called functionalist position defended by Prior, Pargetter, and Jackson (1982).

In this chapter, I will challenge this consensus on the monopoly of microproperties over causal efficacy by showing that macroscopic dispositional properties can be causally efficacious while being distinct from their reduction bases. The importance of this debate lies partly in the impact of its outcome on the conception of mental properties. If the conception of dispositional properties defended here is coherent, then it allows us to think that our desires, beliefs, and other psychological properties give us dispositions to think and act but nevertheless contribute causally to the actions through which these dispositions manifest themselves. I will begin by defending the thesis of the causal efficacy of dispositional properties against a number of very general arguments before turning to the more specific reasons why Armstrong and the functionalists deny that microreducible macroscopic dispositions have causal efficacy of their own.

2. General Arguments against the Efficacy of Dispositions

Of the reasons that have led many philosophers to deny the efficacy of dispositions in general, the most important are the following.

First, dispositional properties are not causally efficacious because they are not observable. I cannot enter here into the controversy over scientific realism: that is, the thesis that theoretical predicates used in science refer to real objects and properties, even when these objects and properties are not directly observable. However, to disarm this argument against the efficacy of dispositions, it is sufficient to note that the impossibility of observing them directly characterizes the dispositional properties of being magnetized, or hard or brittle, in the same way as theoretical properties such as the property of being an electron or that of having a spin of ½. From a realist perspective, unobservable theoretical properties can be causally efficacious. Thus, the mere unobservability of dispositions does not provide any specific reason for denying their efficacy.

Second, another argument is that non-occurring properties cannot be efficacious and that a property cannot be both dispositional and occurring. However, the argument that dispositions are non-occurring is motivated by the following fallacy. Dispositional properties seem to be non-occurring (and therefore incapable of causal efficacy) because they are characterized in a conditional or hypothetical way. The property (or disposition) possessed by any French citizen over the age of eighteen to be potentially, or conditionally, president of the republic is not sufficient to give her the powers of the president. Now all that can be inferred legitimately from this observation is that the disposition does not have the causal powers of its manifestations. This provides no reason to doubt the possibility that the disposition itself gives other powers to its possessors, for example (provided that other conditions are met) to run for president. It is the manifestations of the disposition and not the disposition itself that exist only conditionally or hypothetically. It is fallacious to conclude that a disposition does not occur from the fact that its manifestations do not occur. Therefore, there is no good reason to deny that dispositions are occurrent, even during periods when they do not manifest themselves.[5]

Third, dispositions are permanent states, or static properties, whereas only changes can be causes. There are two ways of interpreting the thesis that mental states do not belong to the logically appropriate category for being causes. According to the first, suggested by Ryle, their temporal permanence prevents them from being causes. I can pass you the salt at the dinner table out of politeness, but politeness is a character trait that characterizes me for a considerable period of time, far beyond the duration of the meal. So my politeness cannot provide a complete causal explanation of my passing you the salt at that moment. The causal explanation of an action — an essentially dated event that takes place at a specific moment in time — requires reference to a dated entity: the cause must be something located temporally and spatially close to the action in question; in short, it must be an event too.[6] The nature of events is controversial, but all conceptions acknowledge that events are located in time. According to Ryle, to find the cause of my act of passing you the salt, we must therefore ask the question "what made him pass the salt at that moment to that neighbour?" (1949, 98). This question cannot be

5 This point was clarified by Martin, who points out that a disposition is "something that is fully real and actual (unlike some of the manifestations).... Dispositions are actual continuants that predate, outlast and may exist entirely without the existence of their manifestations" (1996, 166). See also Mumford (1998, 74).

6 The principle of sufficient reason requires that there be a specific factor at the moment in time when the effect occurs to explain causally why the effect occurred at that moment rather than before or after it. The principle of sufficient reason is valid only in a deterministic framework.

answered by mentioning a mental disposition but only by identifying an event that took place just before the action: "He heard his neighbour ask for it," or "he noticed his neighbour's eye wandering over the table" (98). Dispositions cannot be causes because they lack temporal specificity: causes precede their effects, being contiguous to them in time, whereas a disposition typically exists long before the event that it causes as well as after it.[7]

It might be objected that this argument gives too much importance to the intuitions of common sense. We can accept Mill's (1843) thesis that the distinction between causes that trigger an event and permanent conditions that are less salient but equally necessary for that event is irrelevant to the philosophical characterization of causes. It is undeniable that in most contexts in which we are interested in causes we are more inclined to consider changes rather than stable factors as causes.[8] But from a scientific point of view — and therefore, following Mill, from a philosophical point of view — such stable factors can contribute to an effect just as much as changes. We are therefore justified in regarding them as causes "philosophically speaking" (Mill 1843, Book 3, Chapter 5, para. 3). If I apply electrical voltage to a copper wire, the change in voltage certainly causes a change in the electric current flowing through the wire. But a constant voltage also contributes, along with the resistance of the wire, to the causal determination of the current. Voltage and resistance are stable yet causally efficacious properties of the wire.[9] The static or permanent nature of dispositions shows at most that an event that consists of a change cannot be caused by an event that has only static dispositional properties. But nothing prevents dispositional properties from being causally efficacious.

Two other arguments against the causal efficacy of dispositions have an affinity with the argument that stable states cannot be causes. First, dispositions

7 For the same reason, objects that persist in time cannot be causes: an entity that exists before and/or after the moment that precedes a certain effect cannot be the cause of that effect, although of course it can be a constituent element of a cause appropriately situated in time. See Fales (1990, 54); Steward (1997, 137, 142); Kistler (1999b, 2006a).

8 The pragmatic factors in the context of the explanation and the interests at stake form the basis for the distinction, among the factors that contribute objectively to causing an event, between "the cause" and background factors. See Mackie (1965).

9 "But though we may think proper to give the name of cause to that one condition, the fulfillment of which completes the tale, and brings about the effect without further delay; this condition has really no closer relation to the effect than any of the other conditions has" (Mill 1843, Book 3, Chapter 5, para. 3).

are *facts*, whereas only *events* can be causes. Ryle uses this argument to justify his thesis that the explanation of an action on the basis of mental dispositions, in particular motivations, is never causal. As he puts it, "motives are not happenings and are not therefore of the right type to be causes" (1949, 97). Helen Steward (1997) takes up this argument by pointing out that, when we attribute a disposition to an object, we are referring to a fact, not an event. According to Steward, facts are entities determined by language; for this reason, they lack causal efficacy, although it might be relevant to mention them as part of an *explanation*. This argument has the same source as the previous one, for it is difficult to form event expressions (i.e., expressions referring to events) from dispositional predicates because dispositions are permanent, and it is more natural to conceive of changes than permanent states as events. However, this argument presupposes the Davidsonian conception of the distinction between facts and events, according to which events are particular entities, whereas facts are linguistic entities whose identities are determined by the meanings of the words that designate them (see Davidson 1980). This is not the place to delve into the complex debate on the nature of events and facts (see Kistler 1999a, 1999b, 2006d). I would simply like to point out that the Davidsonian conception has the hardly acceptable consequence of denying the existence of differences, among the properties of a given event, with regard to their contribution to the production of a certain effect event, beyond the pragmatic differences between good and bad explanations. Consider a red billiard ball *R* that hits a white ball *B* at rest with a central elastic shock, so as to set it in motion by transmitting its momentum *M*. Now compare two causal explanations of the fact that *B* has momentum *M* after the shock. The first says that it is the fact that *R* has *M* when it hits *B* with an elastic shock that is causally responsible. The second says that it is the fact that *R* is red when it hits *B* with an elastic shock that is causally responsible. Clearly, the first is not only a good or relevant explanation but also true, and the second is not only a bad explanation but also false. This difference in truth value has its objective basis in the causal relation itself. One way of conceiving of this basis is to say that what makes the former true (its "truth-maker") is the fact that there is a relationship of causal responsibility between the fact that the red ball possesses *M* before the shock and the fact that the white ball possesses *M* after the shock. Conversely, the second (pseudo-)explanation has no truth-maker because there is no relationship of causal responsibility between the fact that the ball in motion is red before the impact and the fact that the white ball has

M after the impact.[10] If this reasoning is correct, then it is legitimate to attribute a causal role to facts, namely that of being terms of the relation of causal responsibility. Thus, the fact that attributions of dispositions normally have a factual, not an eventual, format does not constitute a reason to doubt their capacity to be causally responsible for their manifestations.

Second, Squires (1968) tried to show that the hypothesis according to which dispositions are causes leads to an infinite regress. To explain why a disposition manifests itself on certain occasions but not always, he proposes that it is necessary to postulate the existence of a second disposition: the disposition of the first disposition to manifest itself. Now, of course, this second-order disposition too might or might not manifest itself. Therefore, we need to postulate a third-order disposition to manifest itself and so on. Armstrong (1973, 419) replies by comparing this regress to the infinite series of facts that accompanies any fact *p*: the fact that it is true that *p*, the fact that it is true that it is true that *p*, and so on. We can account for this by distinguishing a linguistic concept of fact, according to which there are indeed an infinite number of different facts, from a "Russellian" concept[11] of a single fact underlying all of these linguistic facts. Or we can, as Armstrong suggests, distinguish between the linguistic expression of a fact and what makes this expression true, its truth-maker. This makes it possible to say that the infinite series of facts accompanying *p* has only one truth-maker, namely *p*. According to the first analysis, there is only one real disposition that corresponds to a unique Russellian fact; according to the second analysis, the unique truth-maker of the infinite series of higher-order dispositions is the first-order disposition, just as *p* is the truth-maker of all of the higher-order facts in the series. Both analyses put us in a position to reject the infinite regress objection by arguing that the apparent infinite series of dispositions

---

10 I have developed this argument for the existence of facts, based on the truth-makers of causal explanations, in Kistler (2002b, 2014).

11 The terminology is due to Bennett (1988, 41). According to Bennett, the identity of a "Fregean" fact is determined by the meaning of the linguistic expression used to express it. The Fregean facts designated by two expressions are identical only if their linguistic expressions can be the subject of an a priori reciprocal derivation, based solely on their meaning (35–37). Conversely, "we sometimes use definite descriptions as though they were Russellian, regarding them merely as pointers to their referents" (39–40). In this sense, two statements can express a single "Russellian" fact even if their meanings are not equivalent (i.e., if the statements cannot be derived from each other a priori).

described by Squires is merely an artifact of language and has only a single fact as its truth-maker, containing a single disposition.

Let us now turn to the most important objection that Molière made famous when he ridiculed the supposed explanatory and causal power of the "dormitive virtue" of opium. Dispositions cannot be among the causes of their manifestations because dispositions are linked to their manifestations by an analytical and therefore necessary relation, whereas causality is a contingent relation. Given that opium has the dispositional property of making people sleep — it possesses dormitive virtue — it seems to be true for purely conceptual reasons that one falls asleep when one has taken opium. It seems that the possession of dormitive virtue cannot be the cause of the fact that the opium smoker falls asleep, because causal relations are always contingent, whereas the conceptual link between a disposition and its manifestation is necessary.[12]

It often happens, at least in my kitchen, that a fragile object breaks after falling on a hard surface. Now it is part of the *meaning* of the predicate *is fragile* that objects to which it applies break when they fall on a hard surface under ordinary conditions. Consequently, the argument continues, since this fragile object fell on a hard surface, the judgment that it broke after it fell is analytical. This implies that, in the sentence "the vase that fell on a hard surface broke because it is fragile," the word *because* designates not a causal relation but a relation of analytical implication based on the meaning of the word *fragile*. So it seems that fragility cannot be one of the causes of the vase's breaking.

The inference to the causal impotence of dispositional properties appears to be valid only if we neglect the fact that the conditionals analytically implied by attributions of dispositional properties are *counterfactual* conditionals and if we forget that the evaluation of these counterfactuals presupposes the determination of a *context*. The judgment "if this fragile vase fell on a hard surface, then it would break," is true only in certain contexts. The *general*

12 Along these lines, Mackie states that "intrinsic powers or specifically dispositional properties in the rationalist sense would violate the principle that there can be no logical connections between distinct existences" (1977, 366), where by "rationalist" he means the doctrine according to which the disposition is not at the same time categorical but purely conditional in the sense that the presence of the disposition makes the manifestation "logically necessary." Mackie then argues against this position on the ground that dispositions are causally efficacious.

conditional, which applies to any fragile object, must make explicit reference to the context in order to be true. "*x* is fragile" implies analytically that, "if *x* falls, under *ordinary conditions*, then *x* will break." Fragility does not manifest itself in all test situations. Even a fragile vase dropped on a hard surface does not necessarily break. We can imagine exceptional circumstances in which the hard surface is mounted on springs and absorbs the shock and others in which the vase and the surface contain powerful magnets that repel each other. These situations are certainly "far-fetched"; however, to show that there is no necessary link (because there is no analytical link) between a fragile object falling onto a hard surface and breaking, it is enough to show that *there are* situations, however unusual, in which the former is not followed by the latter. What follows analytically from the fact that the fragile vase falls on a hard surface is only that it breaks *if the circumstances are otherwise normal*. However, the fact that the fragile vase falls on a hard surface does not imply analytically that it breaks, *tout court*. The fact that it breaks, therefore, remains a contingent fact, and nothing prevents the fragility of the vase from being a factor that contributes causally to its breaking.

Cognitive predicates are dispositional insofar as their meaning can be characterized counterfactually (i.e., by subjunctive conditionals). However, it is not correct to analyze their meanings, following logical behaviourism, with counterfactual statements without a *ceteris paribus* clause. According to Hempel (2002, 17), the statement (*) "Paul has a toothache" has the same meaning as the statement (1) "Paul weeps and makes gestures of such and such kinds," which describes observable behaviour, and with the statement (2) "At the question: 'What is the matter?', Paul utters the words 'I have a toothache,'" which expresses a conditional. Let us admit that it is true that often, or typically, or in ideal or normal circumstances (I will come back to these distinctions later), for a subject *x*, if *x* feels pain (*Fx*), then *x* weeps (*Wx*). But feeling pain is neither necessary nor sufficient for weeping. The fact that Paul feels pain (*Fp*; *F* for feeling pain and *p* for Paul) is not necessary for the fact that Paul weeps (*Wp*): Paul can make gestures "typical" of a toothache because his gums hurt and not his teeth, because he is an excellent actor,[13] or because his motor cortex has been stimulated directly from outside in such a way as to trigger this behaviour, without it being caused by *Fp*. Nor is *Fp*

13 Or a Putnam-style "zombie." See Rey (1997, 153).

sufficient for *Wp*: Paul might hold back and repress the visible signs of his suffering, or ignore the pain, which therefore is not manifest in his behaviour, because he is too busy with an activity occupying his attention.[14] The presence of the dispositional property — and, in (2), of the triggering condition — is not in itself sufficient for its manifestation. For this reason, the link between the dispositional property and its manifestation is not necessary, and the conditional statement that expresses this link is not analytical. There is therefore nothing to prevent the possibility that the fact that Paul feels pain (*Fp*) is causally responsible for the fact that Paul weeps (*Wp*).

It is true that it is analytical to say that, *in general*, fragile things break when they fall or that they tend to break when they fall. This is a consequence of the fact that such subjunctive conditionals express the meanings of dispositional predicates. However, a conditional of this kind is analytical only when it contains the clause "in general." Yet the judgment that a particular glass broke because of its fragility is neither trivial nor analytically implied by the attribution of fragility to the glass. In particular circumstances, a glass might not break when it falls, and it might break for some other reason. The doctrine of the triviality of explanations that mention a dispositional cause rests on a confusion between generic statements such as "the fact that opium possesses a dormitive virtue is causally responsible in general for the fact that the smoker falls asleep" and the particular causal judgment "the fact that the opium that *p* smoked at *t* possesses dormitive virtue was causally responsible for the fact that *p* fell asleep at $t + \Delta t$." The first statement is analytical because it expresses the meaning of the expression "dormitive virtue." But the second statement nevertheless can express a causal relation, instead of being analytical, because dormitivity alone is not a sufficient condition for sleep.[15] Even when it is truly causally efficacious, it produces the effect only in certain circumstances, such as those that prevail for *p* at *t*.

A particularly clear way of showing that the presence of the disposition is not, with the test situation, sufficient for the manifestation is to find possible situations in which the characteristic manifestation does not occur because

14 Putnam (1963) observes that this could be the case systematically; we would then be dealing with "super-spartans" whose conceivability shows that there is no analytical link between the presence of the mental state and the behavioural manifestation that we consider typical.

15 "The causal basis and the striking are not jointly a glass-complete cause of breaking, since the glass does not break" (Bird 1998, 228).

the circumstances contain a factor that acts against the tendency typically produced by the disposition. The absorption of opium by the human organism always produces a tendency to fall asleep, but this tendency does not always result in falling asleep: if *p* has taken the precaution of absorbing an antidote to opium, perhaps an exciting drug, then he might not fall asleep. Goodman gives the example of the attribution of the dispositional predicate "is inflammable" (1983, 39). From the fact that the particular piece of wood *w* is inflammable, one cannot infer the following counterfactual conditional about *w*: "If *w* had been heated enough, it would have burned" (39). This is because it is easy to find situations in which the attribution of the disposition is correct, whereas the conditional is false, for example when *w*'s environment does not contain (enough) oxygen. The presence of the disposition and the triggering factor characteristic of the disposition do not, therefore, comprise a strictly sufficient condition for the existence of the effect. In the case mentioned by Goodman, the effect does not occur in a situation in which a condition not explicitly mentioned, the presence of oxygen, *is lacking*. In other situations, it is the *presence* of an interfering factor that prevents the manifestation from occurring. These two types of situations show that the proposition that attributes a disposition to a particular object does not imply a strict counterfactual conditional about that particular object, linking the triggering to the manifestation.

A distinction must be made between situations in which an interfering factor prevents the manifestation of the disposition from occurring by depriving the object of the disposition and situations in which the disposition remains present (see Mumford 1996, 1998, 86; Bird 1998, 229–30). Examples of situations of the first type include the brittleness of a metal alloy object, which disappears when annealed, or the flammability of wood, which disappears when wet (see Mumford 1998, 86). This type of situation does not call into question the thesis of the implication of a counterfactual by an attribution of disposition to an object, because at the moment when the characteristic manifestation of the disposition does not occur, even though the test condition is present, the object no longer possesses the disposition. The "finkish" dispositions described by Martin (1994) fall into this category. In Martin's thought experiment, a device that Martin calls an "electro-fink" is built in such a way that the disposition never manifests itself when it is triggered. A live electric wire has the disposition to give an electric shock to those who touch it (if their feet touch the ground and if they are not wearing insulated shoes). However, the electro-fink ensures that the disposition disappears if and only if someone touches the wire. It never manifests itself in the characteristic test circumstances. With respect to the test-manifestation pair — <touch the electro-fink electric wire, receive an electric shock> — the circumstances are never "ordinary." Contrary to appearances, Martin's electro-fink is just an extreme case of a perfectly ordinary phenomenon. The electrical circuit in every modern home contains a mechanism whose operating principle is the same as Martin's electro-fink: a differential circuit breaker.[16] To take account of this type of situation, Lewis (1997) suggests requiring that the attribution of a disposition to an object is correct only if the object possesses the causal basis of the disposition for a certain period of time and at least until the moment of the expected manifestation: if x possesses the disposition D to respond with manifestation M to stimulus S, and if Δt is the period of time that elapses between S and M, then, if x is subject at t to S and if x preserves D until $t + \Delta t$, x will manifest M at $t + \Delta t$.[17]

The counterexamples invented by Martin refute the thesis that an attribution of a disposition implies a counterfactual conditional <D & S, M> (where D represents the disposition, S one of D's test conditions, and M the characteristic manifestation of D & S), because in the situations in which a fink is present the disposition disappears in the test situation. But there are also situations in which the disposition remains present yet the characteristic manifestation is absent in a test situation.

There are situations in which a disposition is present in a situation that typically triggers the manifestation but the manifestation does not occur because it is prevented by the presence of an antidote (Bird 1998).[18] An antidote

16 Lewis (1997) and Malzkorn (2000) have attempted to analyze the meanings of attributions of dispositions using counterfactual conditionals that do not use a *ceteris paribus* clause and avoid refutation by cases such as the one imagined by Martin.

17 This formulation differs in several respects from that of Lewis (1997, 157). Lewis assumes, in accordance with the functionalist doctrine, that there is a causal basis B different from the dispositional property D itself, which, together with S, causes the manifestation M. According to functionalism, D itself is causally inert. Lewis also holds that the counterfactual is both necessary and sufficient for the presence of the disposition. Lewis's analysis shows that Mumford (1998, 85) is wrong to judge that the situation described by Martin refutes the thesis that an attribution of a disposition implies the truth of a counterfactual conditional.

18 Johnston (1992) and Molnar (1999, 2003) speak of "masks": they hide the disposition by preventing it from manifesting itself. The existence of such interfering factors is well known, although its importance has been recognized above all in the debate on the existence of strict laws of

is a factor that prevents the characteristic manifestation of a disposition from occurring in a situation in which the triggering factor is present. Let us assume that caffeine is an antidote to opium (Campbell 1860) in the sense that the sleep-inducing effect of opium does not occur when the subject takes both opium and caffeine. Given the existence of caffeine, it is possible for a person to have the dispositional property because of the absorption of opium without it manifesting itself in sleep. This shows two things. First, it is not true to say "for any person *x*, if *x* takes opium, *x* falls asleep," although it is true to say that "for any person *x*, if *x* takes opium in otherwise normal circumstances, *x* falls asleep." Second, given that the implication between the attribution of the dispositional predicate and the manifestation is not analytical, nothing prevents the disposition from causing, in ordinary circumstances, its manifestation. This conclusion applies generally to all dispositions, insofar as there are antidotes for every disposition and every characteristic test situation: that is, situations in which the antidote prevents the manifestation even though the disposition is present. The spring installed under the hard surface, which absorbs the shock, is an antidote to the fragility of the vase.

Here are two other reasons why the link between a dispositional property and its manifestation is not trivial. First, if I explain the breaking of a glass by the fragility of the glass, then I am referring to a property that also has other characteristics apart from leading to breaking when falling. It allows us to characterize the *way in which* the glass breaks: it differs from the way in which the glass breaks if it has the property of explosiveness instead of fragility. Second, the dispositional property is an intrinsic property (Mumford 1998, 138–39). Therefore, attributing causal responsibility for a certain effect to a dispositional property contains the information that the causal responsibility does not belong to the circumstances: a mine placed on the ground would also have the consequence that, if the glass falls, then it breaks, even if the glass is not fragile. But in these circumstances, it would be wrong to attribute the property of being fragile to the glass. Accordingly, when we explain why the glass breaks after falling, saying that its fragility is causally

nature. See Carnap (1956); Stegmüller (1983); Cartwright (1989, 1999). A strict law has no exceptions. The laws of "special" sciences (i.e., sciences other than fundamental physics) are often considered to have exceptions and therefore to be non-strict. Such non-strict laws are said to apply only "other things being equal" or *ceteris paribus*. See Kistler (1999b and 2006d, Chapter 3, 2006b).

responsible informs us that, apart from the fall, the main efficacious property that causes the glass to break is a property of the glass itself.

To express the meaning of a dispositional statement correctly, a conditional statement must include a clause such as "generally" or "in ordinary circumstances." However, it might seem that the presence of such a clause makes the statement tautological. The statement "if *C*, then, in ordinary circumstances, *M*" (where *C* represents a test condition and *M* the corresponding manifestation of the disposition) seems to be equivalent to the tautological statement "if *C*, then *M* or not *M*."[19] There are several hypotheses concerning the "protective" clause that avoid the consequence that any statement containing such a clause is tautological. According to an important proposition (Mumford 1998, 87 ff.), the attribution of disposition *D* to *x* is equivalent to the conditional <*C*, *M*> relativized to *ideal conditions I*:

(*) If *I*, then (if *Cx*, then *Mx*),

where *I* represents the ideal conditions, and both conditionals contained in (*) have subjunctive force (Mumford 1998, 88).[20] The appeal to ideal conditions raises the problem that the context determines what counts as ideal. Take an object that is not fragile, for example a bouncing ball.[21] What counts as ideal depends on the context. Therefore, according to (*), a ball is fragile if there are conditions (which count as ideal in certain situations) such that, if the ball is hit under such conditions, then it will break. Let us take the context of research into the loss of elasticity at extremely low temperatures. Given that the temperature of liquid nitrogen is a context that can be counted as ideal for research on the properties of materials at very low temperatures,

---

19 Several authors (Schiffer 1987; Keil 2000) have expressed doubts about the existence of the referent of such *ceteris paribus* generalizations. According to Martin (1994, 6), for example, the inclusion of a *ceteris paribus* factor in the conditional that serves as analysans makes the counterfactual analysis of the attribution of a disposition trivial. See Lipton (1999); Schrenk (2006). Kistler (2006b) analyzes the role of *ceteris paribus* clauses in taking account of the existence of exceptions to the laws of nature.

20 Bird (1998, 234) suggests analyzing the attribution of a disposition with a counterfactual conditional relativized to "normal conditions," but by this he means what I call here "ideal conditions" (i.e., conditions that are ideal in *relation to a context* of attribution of the disposition).

21 Malzkorn (2000, 459) uses the example of a red rose. The example is not well chosen, because the presupposition that a rose is not a fragile object can be challenged. It is better to use an object that is clearly not fragile.

the ball is brittle according to criterion (*) (see Prior 1985, 5–10; Malzkorn 2000, 459). The problem is that brittleness is attributed to the ball as such, not to the ball in this specific situation. But, intuitively, the ball becomes brittle only under very specific conditions. The fact of mentioning ideal conditions, as in (*), turns out to have the effect of dissociating the disposition from its "normal" effect. (I will come back to the concept of normality in a moment.)

Moreover, if the conditional relativized to ideal conditions is intended to define the *meaning* of the dispositional predicate, then the predicate has as many meanings as there are ideal conditions appropriate to different contexts (see Malzkorn 2000, 459). Most objects would then be fragile in one sense and not fragile in another.

Another strategy is to include the reference to "normal conditions" in the stimulus-manifestation conditional (Spohn 1997; Malzkorn 2000, 457, 459), on the assumption that what counts as normal does not vary according to context in a given world: normal conditions are conditions that occur statistically most of the time.[22] The problem posed by the reference to ideal conditions does not, therefore, arise for the reference to normal conditions: the temperature of liquid nitrogen is not a (statistically) normal condition in relation to the circumstances of the majority of fragility attributions. In this terminology, it is true *a priori* that, "if *x* is immersed in water and the conditions are normal, then *x* dissolves if and only if *x* is soluble in water,"[23] where "normal" has the statistical meaning of "usually, most of the time" (Spohn 1997, 336, 337).[24]

Indeed, it seems to be plausible that we would judge that conditions are not normal when a water-soluble object does not dissolve in water.[25] However,

22 This distinguishes normal conditions from ideal conditions. We can conceive of the initial conditions as being determined by the context in which the disposition is attributed. It can be said (see Spohn 1997) that the concept of normality also has an indexical component, where the index is the whole world and not the particular situation of the utterance; there is only one normal situation in a given world for each stimulus-response relation characteristic of a disposition.

23 "Wenn *x* in Wasser gegeben wird und Normalbedingungen vorliegen, so löst sich *x* genau dann auf, wenn *x* wasserlöslich ist."

24 "gewöhnlich, meistens."

25 I propose to leave aside the following difficulty: what would we say about the conditions of application of the predicate "soluble in water" in a possible world (or an exotic region of the universe of the actual world) where almost all water is saturated with salt? Under normal circumstances in that world (or region), salt crystals do not dissolve when immersed in water. Nevertheless, we would judge that even in such situations salt remains soluble. This means that the reference

this is only plausible in the case of certain paradigmatic dispositions. Molnar (1999, 7) is right to deny that the attribution of a disposition analytically implies a statistical generalization according to which a disposition manifests itself more often than it is masked, on the ground that it "violates the ontological independence of dispositions with respect to their manifestations, by excluding as impossible dispositions that never manifest," and wrongly places "*a priori* constraints on the ratio of responses to stimuli." It can sometimes be justified to postulate a theoretical property rarely manifested. As I will explain in a moment, the analysis of dispositions in terms of normal conditions is still acceptable for common-sense dispositional predicates, but it is not generally acceptable for scientific predicates.

Various authors (Martin 1994, 5–6; Lewis 1997, 157–58; Molnar 1999, 7) have objected that the analysis of dispositional predicates in terms of manifestations produced under normal conditions is as trivial as a *ceteris paribus* law: having the disposition *D* is to do *M* in situations *S ceteris paribus*. In other words, an object has disposition *D* if and only if it does *M* when exposed to *S*, unless it does not. This objection is justified in the case of theoretical properties but not in the case of the semantic analysis of a common-sense predicate: in this case, the statement is not supposed to be nomological, and the reference to normal conditions is not equivalent to a *ceteris paribus* clause. "Normal conditions" are the most frequent conditions in the statistical sense.

3. Dispositional and Theoretical Properties

In order to explain the exceptional behaviour of an object that does not manifest one of its dispositions in a test situation, one can refer to other properties, which might belong to the object itself or to the circumstances. In the clause expressing the dependence of the characteristic manifestation on the test conditions, it is essential to mention the "ordinary circumstances" relevant

---

of the expression "normal conditions" is evaluated referentially and not attributively. According to the attributive mode of evaluation, salt is not soluble in the exotic world in question because solubility is attributed according to the conditions that are normal in the world (or context) where the proposition is evaluated. Conversely, the referential mode of evaluation makes it possible to interpret the meaning of the expression "normal conditions" as what is normal in the context of the utterance and, more specifically, in the immediate environment of the linguistic community. In this interpretation, salt is soluble even if, in the context of evaluation, it is statistically normal for salt not to dissolve in water, as long as it is statistically normal in the context of utterance (i.e., in the actual world, in our immediate environment) for salt to dissolve. See Spohn (1997).

to the case under consideration, insofar as each test situation has an indeterminate number of other properties that can interfere with the manifestation of the disposition.

The scientific conception of properties seeks to get rid of such *ceteris paribus* clauses. In particular, we might look for a scientific explanation of why a disposition *has not* manifested itself in a particular test situation, and in this case the scientific explanation itself must not contain a *ceteris paribus* clause. Let us take the example of a body falling close to the surface of the Earth. If I drop an object supported by nothing else, then it has the disposition to fall a distance of $s = gt^2/2$ in t seconds. However, because of the presence of "antidotes" such as air friction, the disposition will not manifest itself in this way. The discovery of the various antidotes present in a concrete situation explains behaviour that deviates from the direct manifestation of the disposition. This often requires scientific knowledge of properties not directly observable. Ideally, when all of the factors determining the process have been identified, it is possible to explain the behaviour manifested without the need for a *ceteris paribus* clause. This clause expresses our partial ignorance of the circumstances.

Once the scientific description of the situation has been completed, it becomes possible to conceive of dispositional properties as *powers* that necessarily determine their effects.[26] But these effects are not necessarily manifested because they themselves might be powers. For a massive object falling near the surface of the Earth, the scientific conception of the situation makes it possible to substitute for the disposition of the body of mass m to fall $gt^2/2$ metres in t seconds, the force mg that produces a tendency to accelerate with g. Let us call force and acceleration *constraints* or *powers* related according to laws of nature. The force mg determines a power of accelerating with g by virtue of Newton's law, better known by the equation $F = ma$. This tendency to accelerate, although a necessary consequence of the force, does not necessarily manifest itself directly. What is manifested is the result of the superposition (or interaction) of all of the tendencies related to motion. Air friction is another power present in the situation, which imposes on the body another tendency to accelerate in the opposite direction to the first.

26 I will use the term "power" for properties identified by some science: in other words, for natural properties. A dispositional property can be a power in this sense or a property expressed by a predicate in ordinary language, such as "fragile."

We must briefly consider an important objection to the thesis that theoretical properties are powers linked by laws to other properties that, also being powers, do not necessarily manifest themselves directly.[27] Let us say that the law of free fall, considered as a hypothesis, predicts that a body falls $gt^2/2$ metres in t seconds but that observation shows us that the distance of the fall is actually less. My thesis then seems to suggest that this is enough to justify the postulate of a power to fall $gt^2/2$ metres — a power that does not manifest itself directly — rather than taking the observation to refute the hypothetical law of free fall. However, if the disagreement between theoretical prediction and observation was sufficient in itself to justify the postulate of such a power, then it seems that we would justify a general strategy of immunization that would make it possible to justify even the phlogiston theory of combustion.[28] Since observation contradicts the prediction of the phlogiston theory that the residue of combustion has less mass than the body before combustion, my thesis seems to justify the postulate of a tendency, or power, of combustible bodies to become lighter during combustion, a power not directly manifested, however, by a measurable loss of weight. However, we are no longer open to this objection once we impose on the postulate of powers the conditions usually required for the postulates of theoretical entities: the postulate of a power that does not manifest itself directly is scientifically legitimate only insofar as it is possible to give — in each situation in which it does not manifest itself — an independent explanation of the fact that it does not manifest itself.[29] It is legitimate only if it is possible to explain the discrepancy between postulated power and manifestation by the interference of factors whose presence can be detected independently. This means that the Popperian criteria that a hypothesis must satisfy in order to have empirical content and not be ad hoc apply to the hypothesis that explains the discrepancy, as much as to any other scientific hypothesis. The hypothesis of the power of combustible bodies to lose mass during combustion is not legitimate because the only way to reconcile it with the observed fact that the mass of bodies *increases*

27 This objection is analyzed by Lipton (1999) and Schrenk (2006).

28 According to the alchemists, during combustion, combustible substances release a noble substance contained within them called "phlogiston." Ash is the residue that remains once the phlogiston is gone.

29 Pietroski and Rey (1995) show that it is necessary and sufficient to impose such a requirement in order to save *ceteris paribus* laws (i.e., laws that do not apply in all circumstances) from vacuity.

during combustion is to make another postulate that cannot be justified independently: the ad hoc postulate that their mass increases because they give off a substance, phlogiston, whose mass has a negative value. Yet postulating that a body that falls near the surface of the Earth has the power to fall $gt^2/2$ metres in t seconds is legitimate insofar as, in each concrete situation that is the subject of empirical investigation and in which another distance is measured, it is possible to find independent reasons for postulating the existence of interfering factors (which are also powers) — such as air friction — whose superposition on the initial power explains the distance actually observed.

The main conclusion that I propose to draw from this analysis is that the fact of substituting for the disposition to fall $gt^2/2$ metres in t seconds, a force that in turn produces a tendency to accelerate, corresponds to a change of conception of the same property. The first conception of the property involves a dispositional predicate whose meaning is linked to the manifestation; this gives rise to the suspicion of analyticity. Conversely, the second conception of the same property using scientific predicates justifies the idea that it is a real and causally efficacious property. Lawful links between two properties are never known a priori; in other words, laws of nature are discovered a posteriori. Such links are therefore not analytical, which removes any suspicion that properties conceived in this way are "dormitive virtues." Science makes it possible to substitute a categorical conception of a property for its dispositional conception.

The picture that emerges is as follows: the distinction between the dispositional conception of a property and its scientific conception can be based on the following criteria. When we attribute a disposition to an object, we attribute to it a property that exerts a constraint on the evolution of the object, which satisfies three conditions.

First, the dispositional predicate expresses only one of the properties of the object and the situation in which it is found. The attribution of a dispositional predicate is adequate insofar as we *ignore* at least some of the other properties. A fragile object has a property that imposes on it the constraint of breaking when it falls on a hard surface, but properties that we do not know about can impose other constraints on it that act against this first constraint, so as to prevent it from breaking after it falls. In contrast, insofar as the attribution of the property is part of a complete specification of the situation in scientific terms — implying that the outcome is perfectly determined — it is no longer an attribution of a disposition. In a situation in which we know that its tendency to break is counterbalanced by an installation that absorbs the shock, we would not say that the same vase, in this situation, is fragile, insofar as we take account of the whole situation.[30] The hypothesis that some of the other properties of the situation must be unknown for it to be appropriate to attribute a dispositional predicate helps to explain why it is impossible to specify explicitly the "ordinary conditions" under which the disposition manifests itself in a test situation. No such restriction is imposed on a scientific conception of properties. The attribution of a property according to its scientific conception can occur in principle within the framework of a *complete* description of the situation.

Second, the attribution of a dispositional predicate implies the truth of a counterfactual conditional that necessarily contains a *ceteris paribus* clause, whereas the attribution of a scientifically conceived power implies a strict counterfactual conditional without a *ceteris paribus* clause.[31]

Third, we think of a property as dispositional insofar as we think of it as establishing the dependence (*ceteris paribus*) of a manifestation on a test situation, both of which are specified *by observable terms*. Falling and breaking are observable conditions, as are being dropped and falling *s* metres in *t*

30 It is true, however, that the vase itself, independently of the circumstances, remains fragile: it has the capacity to break in other circumstances. Are the sticks of uranium U-235 in a nuclear reactor capable of exploding? Under normal conditions, the boron rods inserted between the uranium rods moderate the chain reaction by absorbing neutrons. The explosion occurs only when the boron rods are removed. Independently of this difference, the uranium rods can be said to have the capacity to cause an explosion. In the presence of the boron rods, the dispositional property is not manifested by an explosion, but when these rods are removed it becomes efficacious and causes an explosion. The fact that its efficacy depends on the context does not show that it is not efficacious; it shows only that its presence is not a sufficient condition for the explosion. Bird explains that "the combination of a [uranium] pile and boron rods . . . does have a disposition to chain-react when the rods are outside the pile, but loses this disposition when they are in the pile. . . . The reactor as a whole . . . , i.e., including the fail-safe mechanism, as long as the mechanism is effective[,] has no disposition to explode at all" (1998, 229–30). The apparent contradiction between this analysis and mine rests on a difference in the choice of object to which the disposition is attributed. The nuclear pile, excluding the boron rods, has the disposition to explode; however, the object composed of both the pile and the boron rods does not.

31 In quantum physics, there are fundamental probabilistic laws; however, these laws make possible predictions of probabilities that are *deterministic* in the sense that these predictions do not depend on partially unknown circumstances, as is the case with predictions based on *ceteris paribus* generalizations, in particular on attributions of dispositions.

seconds.[32] Conversely, the identities of scientifically conceived properties is determined by laws that do not necessarily involve observable properties.[33]

The position outlined here provides an important corrective to Quine's thesis that a disposition is "a partially discerned physical property that will be more fully identified, we hope, as science progresses" (1971, 13) as well as to Armstrong's view that "dispositions . . . are primitive *theoretical concepts*" (1973, 420). My analysis shows that the dispositional conception of a property can *coexist* with its scientific conception. Each obeys its own logic and serves its specific purposes. The distinction between the dispositional and the categorical is epistemic in nature and does not introduce a difference between efficacious and inefficacious properties. Insofar as it is possible to conceive of a property scientifically (i.e., on the basis of the laws of nature that apply to it), it is legitimate to consider it as efficacious, even if it can also be conceived in a dispositional way. The fact that a disposition does not always manifest itself in test situations is explained by the fact that it is not the only property of the situation. Moreover, the fact that such a property is not sufficient in itself to produce a certain effect is no reason to deny that it is causally efficacious. This is also true of clearly efficacious factors, such as the quantity of movement M of the billiard ball mentioned above, causally responsible for the fact that the struck ball has M after the impact, only because the impact is elastic. The uncertainty about the manifestation, expressed in the *ceteris paribus* clause of the conditional linking the test situation to the manifestation, has its origin in the partial ignorance of the circumstances; this is a necessary condition for the attribution of a disposition. Neither the fact that a property is not in

32 The manifestation of certain dispositions, which we might call "spontaneous," does not depend on any particular test situation. Radioactive substances have the disposition to disintegrate, even though no observable factor triggers the manifestation of this disposition. To have a belief is to have the disposition to act as if that belief were true. No external, observable factor is needed to trigger an action that manifests the belief: I can express it by saying a sentence without being prompted by any external stimulus.

33 Theoretical properties have second-order relational properties by virtue of the laws of which they are terms. If a copper cable has the conductivity σ, then this conductivity constrains other properties of the same object by virtue of the laws in which it appears: for example, it constrains the current density and the electric field to be in the ratio $J/E = \sigma$, by virtue of the law $J = \sigma E$. If object o has mass m, and there is another mass m' nearby, then the property of o having mass m places a constraint on o's motion, imposing on it the force $F = Gmm'/r^2$, by virtue of the law of gravitation. The complete set of properties instantiated by the body determines, together with the properties of the environment with which it interacts, its evolution and its causal interactions. For a more elaborate defence of this thesis, see Kistler (2002a).

and of itself sufficient to produce an effect, nor the fact that we do not know whether this effect occurs in a situation that we know only in part, constitutes a reason to deny that it is causally efficacious.

According to the conception suggested here, a property that common sense attributes to objects as a dispositional property can then be scientifically construed as a *power.* The disposition of the opium smoker no longer to feel pain, even in the presence of normally painful stimuli, can be explained in scientific terms: this consists of providing a reductive explanation of analgesia, or insensitivity to pain, a macroscopic dispositional state of the organism, in terms of interactions between certain parts of the organism. In this explanation, the interaction between the morphine molecule and opioid receptors in the superficial part of the dorsal horn of the spinal cord plays a key role. In fact, this area of the body is a strategic zone in the neurological pain circuit, for this is "where the connection is established between the nociceptive fibres and the neurons that transmit the information to the brain" (Besson 1992, 92). The properties of morphine interact with the properties of the body, in this case the opioid receptors in the dorsal horn of the spinal cord, to produce a constraint exerted on the body and thus to produce analgesia. This interaction, and the consequent existence of the tendency to induce analgesia, obey a strict (albeit statistical) law according to which the levogyric enantiomer of morphine binds with a fixed probability to the different opioid receptors. Without exception, the presence of morphine in this area of the body *always* provokes a tendency to induce analgesia. The power of morphine to produce this tendency is always efficacious. However, the analgesic tendency of morphine, the result of an interaction between the properties of morphine and the properties of the body, can be thwarted by another interaction, between morphine and an "antidote," in medical terms, a morphine antagonist, for example naloxone. This molecule prevents the analgesic tendency from producing an observable effect, by binding to opioid receptors, thereby preventing morphine from attaching to them. In the presence of such an antagonist, the tendency of the body that has absorbed morphine not to feel pain any longer is not accomplished.

One attributes the dispositional property of being insensitive to pain to an opium smoker on the mere fact that she has taken opium (and on the fact that she has a human body) *but ignores all of the other properties that her body might possess.* This bracketing of the other properties of the individual to whom a given dispositional property is attributed is at the root of the fact that the dispositional property manifests itself only under ordinary conditions. It explains the indispensability of reference to such normal conditions when making explicit the link between the possession of the dispositional property and its manifestation in test situations. The opium smoker has the disposition to fall asleep. This does not necessarily mean that she will fall asleep, because an antidote can counteract the disposition. But if she does fall asleep, what is the causally responsible property? It might be the dispositional property even if it could have been present without causing the manifestation (it is not sufficient for this manifestation). In other words, designating dormitive virtue as the causally efficacious property is a non-specific way of referring to the causally responsible property.

Something similar happens when we refer to the cause of some event *e* by saying "the cause of *e*." Assuming for the moment that there is only one (complete) cause, the cause of *e* exists, and the expression "the cause of *e*" refers to it. The fact that it can be referred to in this non-specific way does not show that there is no cause of *e*. It exists, and in principle it is possible to identify it more explicitly. But this might require scientific knowledge. In the case of opium, we find that the causally efficacious properties in a case in which the ingestion of opium has caused a person to fall asleep are those chemical properties of morphine whose interaction with the organism produces the tendency to fall asleep. However, we must not confuse the thesis of the existence of causally efficacious properties with the functionalist thesis that "it is the causal basis which is doing the work" (Bird 1998, 233). The disposition is not a property distinct from its causal basis (the former being causally inert, whereas the latter is efficacious). To attribute a disposition to a complex object is to attribute to it (or to its parts) efficacious properties, without being able to identify them precisely. There is only one efficacious property, but there are two ways of attributing it.

The generic statement about what usually happens to opium smokers (and fragile glasses that fall) expresses an analytical link between the possession of the disposition and falling asleep under normal circumstances. Yet the particular causal judgment that such and such a smoker fell asleep because she took opium is not analytical because the link between the possession of the disposition and the manifestation in a test situation is not analytical. "Mrs. *X* smoked opium at *t*" does not allow us to deduce a priori that "Mrs. *X* falls asleep at $t + \Delta t$." However, it might be true that "Mrs. *X* fell asleep at $t + \Delta t$ because she smoked opium at *t*," where the "because" has a causal meaning, in the sense that the judgment affirms not only the existence of a causal relationship between two events but also the responsibility of the fact of having smoked opium for the fact of falling asleep. Why is the generic statement analytical but not the particular judgment? This difference can be explained by the fact that the truth conditions of statements expressing causal responsibility involve properties conceived of in a scientific way. The property causally responsible for falling asleep is a property of the nervous system, which in turn is determined by the chemical properties of the opium absorbed by the body alongside those properties of the nervous system.[34] When we consider its efficacy, we conceive of this property as a power (i.e., as a scientific property) and no longer in a dispositional manner. Causal responsibility is based on the existence of a nomological link between the causally responsible property and falling asleep. The judgment of causal responsibility inherits its a posteriori character from the a posteriori nature of the nomological link. The air of triviality of the explanation of falling asleep following the absorption of opium — by its disposition to induce sleep — disappears as soon as we conceive of this explanation as the outline of a scientific explanation, supposed to indicate the causally responsible fact and the causally efficacious property.

One way of expressing the thesis that a given property can be conceived of both dispositionally and categorically is to say that the dispositional/categorical distinction applies to *predicates* designating properties, or to concepts, but not to the *properties* themselves.[35] The opposite hypothesis can be refuted

34 From a scientific point of view, it is possible to aim for the discovery of the "complete cause" in a given particular case (i.e., the conjunction of all of the properties that interact with the tendency to fall asleep). But even knowledge of the complete cause in a given case does not justify the assertion of strict regularity, because it is impossible to list explicitly all of the factors *absent* in that case that might have interfered. Joseph (1980) points out that the traditional *ceteris paribus* clause would be better called *ceteris absentibus*. Without mention of these absent factors, the presence of all of the factors of the complete cause in some situation does not guarantee that the same effect will occur.

35 Alston was one of the first to challenge "the assumption that the dispositional and 'occurrent' ('episodic') interpretations are incompatible" (1971, 359). The thesis that the dispositional/categorical distinction applies to predicates rather than properties has been defended by Shoemaker (1980), Mumford (1998), and Mellor (2000). "I think that the term 'dispositional' is best employed as a predicate of predicates, not of properties" (Shoemaker 1980, 211). "Dispositionality is a feature not of properties but of predicates, namely of those whose application conditions can be stated in reduction sentences. . . . Properties in our sense . . . need not in themselves be either dispositional or categorical; those that exist can just be" (Mellor 2000, 767–68). Lowe distinguishes between "'occurrent' predication" and "'dispositional' predication" (2001a, 11). However, his view is

in the following way. Suppose that the dispositional/categorical distinction applies to properties themselves, independently of the predicates that we use to refer to them. All natural properties are components of laws of nature. It is by virtue of such laws that the objects possessing natural properties also have other specific natural properties. This is also true of properties that we do not intuitively regard as dispositions, such as the property of a gas to have a certain temperature T or the property of a stone to have a certain mass. According to the ideal gas law, all (ideal) gases that have temperature T also have, in volume V, pressure $p = nRT/V$ (where R represents a constant factor and n indicates the quantity in moles of gas molecules contained in the sample). Now the existence of this law provides us with a way to conceive of properties as dispositional: it gives its possessor the disposition to have another property linked to it by the law. Temperature is a dispositional property insofar as it gives the gases that possess it the disposition to have pressure p when they occupy volume V. Similarly, the fact of having a certain mass gives a stone — thanks to the law of gravitational attraction — the disposition to move toward other massive bodies and, in particular, to fall when close to the Earth's surface. Therefore, the hypothesis that dispositionality is a property of properties, and not a property of predicates or of our concepts of properties, leads to the result that *all* natural properties — that is, all properties that appear in laws of nature — are dispositional. But this is clearly incompatible with our intuitive understanding of the concept of disposition according to which not all properties are dispositional.

The Popperian thesis that all properties are dispositional seems to oppose this intuition and thus to undermine my argument.[36] However, it can be interpreted in a way that is compatible with my thesis that the dispositional/

incompatible with mine: according to Lowe, dispositional predication ascribes a universal property to an object via a kind of object of which it is an instance, whereas occurrent predication ascribes an instance of a property to the object. This distinction cannot account for the difference between the dispositional and the scientific attribution of a property, insofar as it does not involve the semantic link between the disposition and its manifestation characteristic of dispositional predicates. Lowe conceives of the distinction in purely ontological terms; however, ontologically, the same property is involved in both kinds of attribution: the instance that is the object of an "occurrent predication" is an instance of the same universal property that is the object of a "dispositional predication."

36 See Popper (1957). This thesis has also been defended or at least suggested by Harré and Madden (1975), Thompson (1988), Cartwright (1989), Blackburn (1990), and Harré (1997). Goodman (1983) puts forward the more moderate thesis that there are many more dispositional predicates than appear at first sight.

categorical distinction concerns our *concepts* of properties rather than the properties themselves. We have just seen that there is a way of conceiving of any natural property dispositionally. If *P* is any natural property linked by a causal law to another property *R*, then knowledge of this law enables us to conceive of *P* as "the disposition to cause *R*." In this interpretation, Popper's thesis loses the counterintuitive character that we usually attribute to it when we interpret it to mean that all properties are dispositional and, consequently, that no properties are categorical.[37] As long as the distinction is conceived of as relating to concepts or predicates, there is nothing to prevent the same property from being conceived of alternatively as a dispositional property and a categorical property.[38]

A promising way of conceiving the categorical/dispositional distinction at the level of predicates was suggested by Shoemaker (1980).[39] A predicate *D* is dispositional if and only if its attribution entails *analytically* — that is,

37 There is a formidable objection to this interpretation of the thesis, variants of which have been put forward by Holt (1976), Robinson (1982), Blackburn (1990), Armstrong (1999), and others. However, this objection that the thesis that all properties are dispositional makes us "lose the substance of the world" (Holt 1976, 23) does not apply to the interpretation suggested here.

38 Here are two other ways of reconciling the fact that most properties can be thought of dispositionally with the paradoxical appearance of the thesis that all properties are dispositional. First, according to Martin (1996) and Heil (2004), every property has a "dual nature": that is, it "endows its possessor with both a particular disposition or 'causal power' and a particular quality" (Heil 2004, 197). However, rather than providing a solution, this is simply a way of posing the problem of understanding how these two apparently incompatible "aspects" can nevertheless coexist. Second, to avoid the conclusion that all properties are dispositional, while considering that the categorical/dispositional distinction applies to the properties themselves, Jackson introduces a distinction between properties bound *essentially* to a certain manifestation in certain circumstances and others thus bound only *contingently*. "What makes a property a disposition is that it itself is *essentially* linked to the production of certain results in certain circumstances" (Jackson 1998, 101). However, insofar as the link between a natural property and its manifestation in characteristic circumstances is based on a law, it is difficult to justify this distinction. It presupposes that there is, among the laws in which a property appears, a first set of laws essential to the property and a second set of laws that apply to it only contingently. The property could exist even if the laws in the second set did not exist, whereas it could not exist without the laws in the first set. The distinction between two sets of laws, in terms of the modal force with which they determine the identity of the property, would require an independent justification. In Kistler (2002a), I give arguments in favour of the opposite thesis, according to which all of the laws in which a property appears are essential to it.

39 Shoemaker illustrates his analysis with the example of the predicate "is made of copper," which is "not dispositional in this sense. There are causal powers associated with being made of copper — for example, being an electrical conductor. But presumably this association is not incorporated into the meaning of the term 'copper'" (1980, 210). The distinction that I offer here is not the same as Mumford's (1998), insofar as I define it in terms of the a priori/a posteriori distinction, whereas

by virtue of the meaning of the predicate alone — a counterfactual linking a test situation to a characteristic manifestation. The statement "this vase is fragile" implies analytically that, if the vase were to fall under ordinary circumstances onto a hard surface, then it would break. Conversely, the statement "this vase is made of fine clay" implies the same counterfactual but not analytically. In this case, the implication is based not on the mere meaning of the predicate "to be made of fine clay" but on laws known only a posteriori. The laws in which a natural property participates guarantee the existence of such counterfactuals, but their knowledge is not always part of the meaning of the predicates with which we refer to these properties. This difference is the basis for the distinction between dispositional predicates and categorical predicates. The attribution of a dispositional predicate implies the counterfactual that links a test situation to a manifestation analytically and therefore a priori, whereas this implication is a posteriori in the case of categorical predicates.

4. The Epiphenomenal Trilemma of Macroscopic Dispositions

The reality of *macroscopic* dispositional properties is often questioned for reasons to do with their relationships with underlying microproperties. The concept of causal basis plays a key role in such arguments. All dispositions have manifestations. The causal basis of a disposition is what causes its manifestations. From this conception, it follows that all dispositions have causal bases since something must cause their manifestations. Insofar as they are efficacious, the constituent properties of such a basis are categorical properties. However, in the case of macroscopic dispositional properties, there are several ways of conceiving of the relationship between a disposition and its causal basis that could lead one to deny that the disposition has causal powers of its own.

First, according to the "functionalist" conception of dispositions (Prior, Pargetter, and Jackson 1982), a disposition is a second-order property. However, only first-order properties can be causally efficacious. There are two reasons for thinking that dispositions are not first-order properties: one

---

Mumford sometimes characterizes it by saying (like Jackson 1998) that categorical properties are *contingently* linked to their nomological consequences.

disposition can have several different bases, and a disposition has its base (or bases) only contingently (one disposition could have other bases than it actually has).

Second, according to an important concept of reduction (Causey 1977), the discovery of the reduction of a macroscopic property to microscopic properties is the discovery of an identity. Insofar as a dispositional property is identical to its categorical reduction basis, it has no proper causal efficacy beyond that of its basis (Armstrong 1973).

Third, the concept of functional reduction offered by Kim (1998) brings together elements of the first two conceptions. According to this third conception, to attribute a disposition to an object is to attribute a second-order *predicate* to it: the object has a property that plays the role of causing the manifestation (under ordinary circumstances) under test conditions. The predicate specifying the role is second order, insofar as the reference to a property that plays the role is equivalent to an existential quantification over first-order properties. Only the property that plays the role is causally efficacious; the disposition that corresponds to the role is not. However, according to Kim's conception, the property playing the role is necessarily microscopic, even when the disposition is attributed to a macroscopic object.

None of these conceptions accepts that macroscopic properties have their own causal efficacy. This constitutes a dilemma, more precisely a trilemma, insofar as there is apparently no other possibility and none of these possibilities seems to be compatible with the common-sense intuition that our cognitive properties are at the causal origins of our actions, without being identical to any microscopic property of our brain.[40] I call it the "epiphenomenalist" trilemma because all of the alternatives deny that macroscopic dispositional properties have their own causal powers. Insofar as they have identities of their own, they are epiphenomenal. The first horn of the trilemma consists of considering dispositions as inefficacious, and the second and third horns

40 The question of whether this intuition is correct is controversial of course. It is the subject of a now classic debate between Wittgenstein (1958), who contests the coherence of an entity (a "mental representation") whose content justifies an action and at the same time is causally responsible for that action, and Davidson (1963), who argues contrarily that it is necessary for our conception of ourselves as agents who exercise causal power over our own actions to suppose that our reasons for acting are simultaneously the causes of our actions. The intuition that I am talking about is compatible with Davidson's position but not with Wittgenstein's. See Kistler (2006c).

consider them to be efficacious only insofar as they are identical to their microscopic categorical bases.

However, the trilemma can be avoided by conceiving of macroscopic dispositions as efficacious properties not identical to their microscopic bases, if two premises are accepted:

(1) the dispositional/categorical distinction applies to predicates, not to properties, and

(2) the categorical basis is not necessarily the reduction basis.

Rejecting these two theses leads to the first horn represented by functionalism, and accepting (1) but not (2) leads to the last two horns of the epiphenomenalist trilemma, represented by Armstrong's and Kim's conceptions of dispositions.

Armstrong agrees with thesis (1) that dispositionality and categoricality are two ways that one may conceive of properties, in themselves neither dispositional nor categorical. However, he denies that there are macroscopic dispositional properties distinct from the microscopic properties to which they are eventually reducible. Armstrong takes the example of an occurrent fragile state of a piece of glass. This state can be causally efficacious when it contributes to making the glass break when it falls. When we do not know the causally efficacious properties that are intrinsic properties of the glass, we refer to them with a definite description, in terms of their typical effects. The predicate *is fragile* is defined by typical effects that occur in ordinary circumstances. However, we are dealing here with only two ways of referring to a single state, one direct (but inaccessible to us because of our ignorance of the intrinsic nature of the property), the other indirect and referring to its causes and effects. Two ways of referring to a state do not transform it into two states or two properties. This difference is the result of "a verbal distinction between the disposition and the state. (A verbal distinction that cuts no ontological ice.)" (Armstrong 1973, 419). According to Armstrong, disposition is a concept that corresponds to a certain functional way of referring to properties or states rather than to a particular kind of property or state. "Dispositions . . . are marked off from (many) other states by the way they are *identified*" (419).

So far, Armstrong's view is compatible with my thesis (1). The controversial step in his reasoning is the following. Armstrong considers that the only way to make sense of the idea that the same property can be conceived of, either dispositionally or categorically, is to suppose that this property is actually (identical to) its microscopic basis of reduction, which he calls the "categorical basis." However, Armstrong gives no justification for the implicit premise that only a microscopic property can be categorical and efficacious.[41] As he puts it, "what then is the disposition, the brittleness? It is the 'categorical base,' the microstructure, but it is this property of the object picked out not *via* its intrinsic nature, but rather *via* its causal role in bringing about the manifestation" (1996, 39).[42] In the case of brittleness, it is a property of the chemical bond between the molecules composing the glass. In the case of the disposition to transmit hereditary characteristics, it is the microscopic properties of DNA molecules.

> A good model for the identity of brittleness with a certain microstructure of the brittle thing is the identity of genes with (sections of) DNA molecules. Genes are, by definition, those entities which play the primary causal role in the transmission and reproduction of hereditary characteristics. . . . In fact sections of DNA play that role. So genes are (identical with) sections of DNA. (Armstrong 1996, 39).[43]

---

41 This premise is often accepted without argument. Mackie characterizes his own "realist view" of dispositions by saying that "there will always be an occurrent ground" and immediately moving on to assert that this occurrent ground is necessarily different from the dispositional property itself. "This ground will not in itself be specifically dispositional" (1977, 365), his example being the categorical microproperties underlying the macroscopic disposition of the solubility of sugar in water. "In crystalline sugar the feature causally relevant to its solubility in water will be something about the bonds between the molecules in the crystal structure" (365).

42 Armstrong starts from Quine's thesis that, by the use of a dispositional predicate, "we can refer to a hypothetical state or mechanism that we do not yet understand" (Quine 1971, 10), a use that can be replaced by a direct way of referring to it as soon as science has discovered the intrinsic nature of this state. A disposition, according to Quine, is "a partially discerned physical property that will be more fully identified, we hope, as science progresses" (13). See Armstrong (1968, 86; 1997, 73).

43 See also Armstrong (1968, 90). He encounters a difficulty because, on the one hand, he maintains that the *truth-making* relation (between a proposition and a state of affairs) is necessary, so that, if having a certain molecular structure makes the attribution of the disposition of being fragile true, then it is necessary that all things that have this molecular structure be fragile. On the other hand, he holds that laws are contingent and that the relationship between having the molecular structure and breaking after falling depends on laws. Later Armstrong (1997) adopts a different position, according to which what makes the attribution of fragility true are both molecular structure and laws.

In the case of dispositional mental properties, the underlying properties are microscopic properties of the brain.[44] Armstrong offers no reason to think that the categorical basis of a disposition is necessarily microscopic. However, the debate in which he develops his theory suggests a hypothesis about the origin of this thesis. In the passage preceding the introduction of his thesis on dispositions, Armstrong criticizes Ryle (1949) and Price (1953) for two mistakes. The first is their rejection — via a priori reasoning — of the search for a microscopic basis that would allow the reduction of a given macroscopic dispositional property, in particular a mental property. The second is the verificationist mistake of refusing to accept the existence of theoretical properties whose identity conditions are independent of any particular verification procedure.

Armstrong's reasoning would be valid if these two mistakes were only one mistake. If that was the case, then it would be natural to think that a single manoeuvre is necessary and sufficient to avoid the mistake; however, the only way to avoid the two mistakes in one step is to postulate a theoretical property underlying the disposition that at the same time is the microscopic property that provides the basis for its reduction. The identification of the disposition with a microscopic theoretical property avoids the verificationist mistake since it is a theoretical property whose identity is independent of any particular verification procedure, and it avoids the mistake of a priori denying the possibility of a microreduction.

However, there is no reason to believe that Ryle and Price committed a single mistake. To avoid the mistake of verificationism, it is necessary and sufficient to postulate a categorically occurring theoretical property whose identity is not reduced to a pair <test condition, manifestation> but might contribute in different and complex ways to different causal processes. Now there is nothing to prevent the property thus postulated from being a *macroscopic* property belonging to the same object as the disposition: the person — and not some of her neurons or neuronal circuits — in the case of mental properties and the body — and not some section of its DNA molecules — in

44 Armstrong (1968, 76–77) develops the analogy between these two identifications with regard to their contingency. However, Armstrong (1997, 73) explains that this contingency arises only from the contingency of the laws of nature that cause DNA, by virtue of its properties, to play the role of genes and that cause the brain, by virtue of its properties, to play the roles that characterize mental states.

the case of the ability to transmit hereditary characteristics. Therefore, it is conceivable to remedy verificationism, without at the same time remedying the other mistake, which consists of rejecting the possibility of microreduction. This second mistake can be avoided by a second step independent of the first step. Microreduction of a macroproperty consists of the discovery of a nomological explanation of the existence of the macroproperty of an object, based on the microproperties of its parts and their interactions by virtue of laws. Molecular biology makes it possible to explain, on the basis of the microscopic components of the human body and their many complex interactions, how organisms can pass on some of their hereditary characteristics to their offspring. However, this reduction, assuming that it is complete, does not justify the identification of the dispositional property of having the capacity to transmit hereditary characteristics to any particular microscopic property. In particular, this capacity might not be identical with any microscopic property of (sections of) DNA molecules.[45]

Once we have distinguished the two steps that separate the dispositional conception of a macroproperty from the discovery of the microscopic basis for its reduction, it becomes clear that the expressions "causal basis" and "categorical basis" are used in two fundamentally different senses: according to their first meaning, they designate the categorical property underlying a disposition causally responsible for its manifestations, a property that might well be macroscopic (i.e., it might belong to the same object as the disposition rather than to its parts). Their second meaning is strongly suggested by the term "base": when a microreduction of the macroscopic property underlying the disposition has been discovered, what is commonly referred to as the "basis of reduction" is the set of microscopic properties of the parts of the object that determine the macroscopic property. Armstrong's mistake is in

45 See Kitcher (1984); Rosenberg (1985); Morange (1998). Armstrong is well aware that identifying a gene with a segment of DNA is an oversimplification; however, he thinks that this does not threaten the coherence of his position: "The statement 'The gene is the DNA molecule' is not a very exact one from the biological point of view. But it will prove to be a useful example in the development of the argument, and it is accurate enough for our purposes here" (1968, 77). It seems to me, however, that what is at stake here is not a matter of neglecting some details: it is a fundamental mistake to take a microscopic property of a part of the organism as the causal basis of the organism's disposition to transmit its hereditary traits. The causal basis of hereditary transmission consists of a complex mechanism of which DNA is only one part. The property of possessing this mechanism can be attributed only to the organism as a whole and not to one of its microscopic parts, be it DNA.

confusing these two meanings of "basis" (or "base") and in admitting without justification that the categorical basis in the first sense of the term must necessarily be the basis in the second sense of microreduction base.

Here is a reason to think that there are at least some categorical bases that are not reduction bases. Let us suppose that the hierarchy of levels of composition of macroscopic objects is not infinite but contains a level of absolutely "atomic" objects and properties no longer microreducible because the objects belonging to this level have no parts. In this case, the chain of microreductions stops with the discovery of this fundamental level. A property *M* belonging to this absolutely fundamental level gives causal powers, at least indirectly, to its possessors; otherwise, there would be no reason to postulate its existence. The powers that *M* gives to its possessors have a categorical basis: the properties of the object causally responsible for the manifestations of those powers. Now, given that *M* has no reduction basis, this categorical basis can consist only of *M* itself or of other properties belonging to the level of *M*.

My thesis that dispositionally conceived properties can also be categorically conceived suggests a simple solution to the "problem of the missing reduction base" raised by Molnar (1999, 8). Molnar shows the implausibility of three attempted solutions to this problem, created by the difficulty to accept both of the following two theses. Every disposition has a categorical basis, in the sense of a microreduction basis, and there are fundamental particles that have absolutely no structure. The three rejected attempts to reconcile the two are the following. (1) The causal basis of particles without internal structure consists of global properties of the universe.[46] (2) It is speculated that an infinite hierarchy of levels of structure is hidden beneath the level that appears to us as absolutely fundamental. (3) The fundamental properties of absolutely simple particles are not dispositional. After rejecting these three propositions, Molnar concludes that, "when it comes to the fundamental micro-entities, no suitable properties exist that could serve as a causal base for their dispositions" (17). Similarly, Mumford concludes that, in the case of a fundamental property for which there is no microreduction, "we have . . . just the one mode of characterising it available to us, the dispositional" (1998, 169).

This conclusion is paradoxical insofar as it implies that the manifestations of the powers provided by fundamental properties have no cause. My

46 Harré (1986, 295) has proposed the idea of such an "ultra-grounding."

distinction concerning the two meanings of the term "basis" offers a way of avoiding this: in the absence of a microreduction basis, the way to avoid this conclusion is to suppose that the dispositional property of a fundamental particle is itself the categorical basis causally responsible for its manifestations.[47]

Let us see how my two theses (the dispositional/categorical distinction applies to predicates and not to properties, and the categorical basis is not necessarily the reduction basis) make it possible to avoid the conclusion of the so-called functionalist conception of dispositions according to which macroscopic dispositional properties are epiphenomenal (Prior, Pargetter, and Jackson 1982), where causal efficacy is reserved for microscopic bases.

According to Prior, Pargetter, and Jackson, if a dispositional property is designated by a second-order macroscopic predicate, then its categorical basis is a first-order microscopic property. Prior, Pargetter, and Jackson put forward two reasons to justify the "distinctness thesis" (1982, 253), according to which the causal basis is neither dispositional nor macroscopic. The first reason is that a disposition typically has several bases. This prevents, by virtue of the transitivity of identity, the identification of each of them with the dispositional property. The second reason is that, even in the case of dispositions that have only one causal basis, the disposition has its basis only contingently, whereas the identity of a dispositional property with its basis should be necessary, given that the corresponding predicates are rigid designators (Kripke 1972). However, the conjunction of the distinctness thesis and the thesis that the causal basis causes the manifestations of the dispositional property implies that the disposition itself is causally inert. If the basis causes the manifestations, along with the triggering situation, and if the disposition is distinct from the basis, then, if generalized overdetermination is excluded, the dispositional property itself is causally inert.[48]

---

47 Hypothetical dispositions for which there is no microreduction are often called "ungrounded dispositions" (see Mumford 1998, 167; Molnar 1999). However, in the sense of "basis," where this expression designates the property of the possessor of the disposition causally responsible for its manifestations, it is clear that every disposition necessarily has a basis. Therefore, to say that a disposition has no basis can only mean that it cannot be reduced and therefore has no *reduction basis*.

48 Systematic but accidental overdetermination is implausible. It is implausible that every effect has several complete causes that act in parallel but independently of each other. However, it is not implausible that there can be two sets of properties instantiated at the same place and time, both sufficient for the same effect, that *are not independent* insofar as one set of properties nomologically determines the other. See Chapter 5.

The functionalist view makes two mistakes.[49] It is correct that a predicate that defines a property dispositionally (or functionally) is a second-order predicate, insofar as it contains an existential quantification over the properties that play the role. It is also correct that, insofar as a property is causally efficacious, it can be conceived of with a first-order concept. Yet it is incorrect that the existence of two ways of conceiving of a property entails the existence of two properties and to infer from the fact that the efficacious property is expressed by a first-order predicate to the fact that the property is microscopic.

The distinction between the first and second orders that concerns predicates and concepts is independent of the distinction between the macroscopic and the microscopic, which concerns properties (see Kim 1997b, 1998). Given that functionalists neglect these two distinctions, they arrive at the following three mistaken conclusions: (1) the "causal basis" is necessarily microscopic, (2) dispositional properties — rather than the predicates that express them — are of second order, and (3) therefore they lack causal efficacy.[50]

5. The Example of Colour Representation

The perception of colours by the human visual system can serve as an illustration of the argument developed above. In psychology and psychophysics, colour representations are conceived of as macroscopic dispositional properties of persons: they give them the disposition to make similarity and discriminability judgments.[51] To explain these judgments, which are observable data, we can postulate the existence of a mental or cognitive space specific to colour representation. Such a postulate goes hand in hand with the construction of a theory that describes unobservable entities in such a way as to be able to explain a certain number of observable facts and empirical regularities. Shepard (1962) showed that similarity judgments[52] contain sufficient constraints to de-

49 I found the same mistakes in the a priori reduction of Jackson and Chalmers (see Chapter 2). See Kistler (2004b, 2005d).

50 Armstrong avoids the mistake of inferring from a duality of conceptions to the existence of two properties, but he makes the functionalists' mistake of thinking that efficacious first-order properties are necessarily microscopic. See Block (1990); Armstrong (1996).

51 Two colours are discriminable if a normal subject can distinguish them systematically under ordinary conditions.

52 See Clark (1993). Shepard uses only the order of similarity between the pairs of stimuli presented to the subjects, as it appears in the subjective judgments of subjects, without using quantitative estimates made by the subjects of the apparent distances between the stimuli. In

termine,[53] for any domain of perceptual qualities, (1) the minimum number of dimensions that the cognitive space must have in order to represent the stimulus domain in question, and (2) the location, in the cognitive space, of each represented stimulus (i.e., the coordinates of the representations of the different stimuli in the cognitive space). The algorithm developed by Shepard enables him to construct "maps" of a number of cognitive spaces corresponding to different stimulus domains: the "structure of proximity" by which a subject represents, among other things, the different facial expressions of other members of his species (Shepard 1962), colours (Shepard 1962, 1965), the consonants of his mother tongue (Shepard 1974), musical intervals (Shepard 1974), and familiar animals (Shepard 1974).

In the case of colours, the first result obtained by the algorithm is that two is the minimum number of dimensions that the cognitive space containing the colour representations visible to the human visual system must have (leaving aside the dimension of luminosity). Any assumption of a simpler cognitive structure would be incompatible with the experimental data. It is impossible to account for judgments of similarity between (perceptions of) colours on the basis of a cognitive space of representation that has only one dimension: on the basis of judgments of similarity between red and yellow, yellow and green, green and blue, blue and violet, we could attempt to situate the representations of these colours within a single dimension in the order of the rainbow. However, this one-dimensional representation would not account for the perceptual similarity between red and violet. If the representation of colours were structured in one dimension in the order of the rainbow,

addition to direct judgments of similarity, Shepard uses data obtained by more indirect methods that make it possible to judge the proximity of stimulus representations in cognitive space. In particular, these methods use the frequency with which subjects confuse different stimuli, the delay required to discriminate between two stimuli, or (for small children and animals) the size of the orientation reflex when the first stimulus is replaced by the second.

53 Mathematically, Shepard's algorithm uses two a priori constraints. First, it assumes that the function relating apparent similarity to proximity in the representation space is *monotonic*. The monotonicity of the function guarantees that, if colours *A* and *B* are judged to be more similar than colours *C* and *D*, then the distance, in cognitive space, between representations *R*(*A*) and *R*(*B*) is less than the distance between representations *R*(*C*) and *R*(*D*), and conversely apparent similarity depends, in the same systematic way, on the distances between representations. In particular, the representations of the stimuli judged to be the most similar must be separated by the shortest distance. Second, cognitive space has the smallest dimension that allows us to construct a monotonic and unique function linking apparent similarities to distances between representations in cognitive space.

then red and violet would have to be the most dissimilar colours, whereas in reality they are more similar to each other than each is to green or yellow, for example. Shepard's second result is that there is a unique topological structure or "map" of represented colours related by a monotonic function to similarity judgments. In Figure 3.1, the representations of the different visible colours are located on a circle.

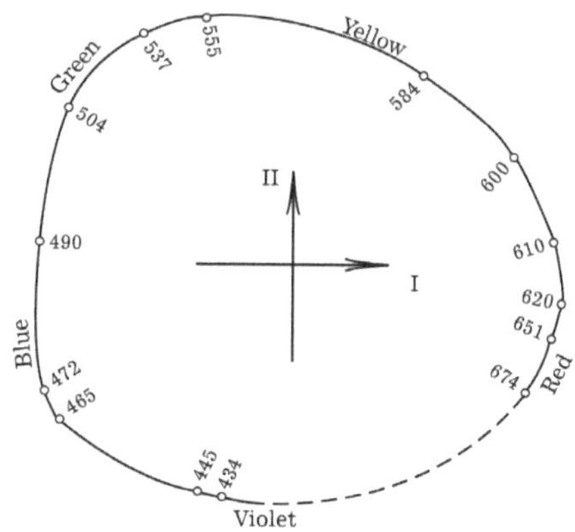

Figure 3.1 Representations of some colours in cognitive space. The wavelength of the colour stimuli is indicated in nm (nanometres). Adapted from Shepard (1962, 236).

Insofar as colour representations are conceived of as intermediaries between stimuli and similarity judgments, they are indeed dispositional properties. However, once we have freed ourselves from the verificationist prohibition that prevented Ryle from conceiving of representations as entities independent of any particular manifestation, we can consider colour representations as theoretical entities that make it possible to produce causal explanations of similarity judgments. As theoretical entities not directly observable but postulated in order to explain observable phenomena, they belong to the same category of entities as protons and neutrons, whose existence makes it possible to construct causal explanations of phenomena observed after the interactions of particles produced in a particle accelerator.

The hypothesis of the existence of colour representations and of the structuring of these representations in a cognitive space is independent of the discovery that these representations and this space can be *reduced* to neurophysiological entities. Similarly, the legitimacy of the postulate of the existence of protons and neutrons is not conditioned by their possible microreduction, an issue independent of their existence. The neurophysiological reduction of colour representation is the subject of intense research, which suggests the existence of a complex mechanism involving several areas of the cortex, in particular area V1 of the visual cortex and the inferior temporal cortex (Conway et al. 2010). Such a reduction has already been achieved in the case of a number of other cognitive spaces, corresponding to the representation of certain perceptual domains in certain animals: the neural structure used by the barn owl (*Tyto alba*) to represent the location of a sound source has been identified in the upper layer of the optic roof of this animal's brain (see Gallistel 1990, 478 ff.); neuroscientists have also succeeded in discovering the neural structure located in the auditory cortex of the bat used to represent the position and speed of an object using echolocation (see Gallistel 1990, 492 ff.). In each case, the subjective sensation — which causes the action or judgment — results from the simultaneous activation of a large number of neurons located in the area corresponding to the cognitive map of the relevant perceptual domain. According to one hypothesis, the representation produces its effects through a mechanism equivalent to the vector calculation of the average activation, performed on all of the neurons in the relevant area (see Gallistel 1990, 489, 515; Churchland and Sejnowski 1992, 233–37).

The lesson that I propose to draw from this brief examination of some of the results of psychological and neurophysiological research on the representation of colours is that it is coherent to conceive of the categorical basis of the psychological property that produces the manifestations of the representation of colours as a theoretical macroscopic property. It belongs to the person rather than to its microscopic parts, such as its neurons. It is an independent issue whether a microscopic (i.e., neuronal) basis can be found that provides a reductive explanation of this macroscopic psychological property.

6. Dispositional Properties with Multiple Manifestations

An important reason for considering that there are powers (i.e., real properties that make dispositional statements true) is that a natural property generally makes true a whole set of attributions of dispositions.[54] This is often

54 The reasoning set out in this section is developed in Kistler (2012, 2020).

expressed by saying that many dispositional properties are "multi-track" (i.e., they can manifest themselves in several ways).[55]

Let us take the example of an electric charge. The fact that an object x has the elementary electric charge q makes true the attribution of several dispositions to this object:

1. The disposition of x to undergo a force $\vec{F} = q\vec{E}$ if x is in the electric field $\vec{E}$.

2. X's disposition to attract a second object carrying a charge q^* at distance r with a force of $F = k_e \, {qq^*}/{r^2}$ (Coulomb force), where k_e is a constant.

3. The disposition of x to undergo a force $\vec{F} = q\vec{v} \times \vec{B}$ (Lorentz force) if x moves with speed $\vec{v}$ in a magnetic field $\vec{B}$.

4. The disposition of x to carry a magnetic moment $\vec{\mu} = {1}/{2}\, q\vec{r} \times \vec{v}$ if x is rotating with speed $\vec{v}$ around a circle of radius $\vec{r}$.

These four dispositions are not identical: it is not the same thing for the object x to have the disposition to undergo a force $\vec{F} = q\vec{E}$ (manifestation M_1), if x is in an electric field $\vec{E}$ (test condition T_1), to have the disposition to exert the force $F = k_e \, {qq^*}/{r^2}$ (manifestation M_2) on a second object located at distance r and carrying an electric charge q^* of the opposite sign (test condition T_2), and to have the disposition to undergo the force $\vec{F} = q\vec{v} \times \vec{B}$ (manifestation M_3), if x moves with speed $\vec{v}$ in a magnetic field $\vec{B}$ (test condition T_3).

The fact that these dispositions are different might seem to be paradoxical, insofar as they are all dispositions of bearing the elementary electric charge, a single property. The paradox can be avoided by distinguishing these dispositions from the underlying natural property that I have called a "power." The postulate of such an underlying natural property is a way of accounting for

55 Ryle envisages the possibility of dispositions that possess an infinite number of possible manifestations. "The higher-grade dispositions of people . . . are, in general, not single-track dispositions, but dispositions the exercises of which are indefinitely-heterogeneous" (1949, 32). See also Mellor (2000, 760).

the fact that the dispositions mentioned are indissociable: nothing has one of these dispositions, or part of them, without having them all.

What is the relationship between the property of being electrically charged and the various dispositions D_i that this property gives to a charged object? The property of being charged cannot be identical to all of these dispositions D_i, insofar as the D_i are not identical to each other. Nor can it be identical to one of them to the exclusion of the others, insofar as it is not more closely linked to one of the D_i than to the others.[56]

The concepts of natural property and power are metaphysical, whereas the concept of disposition is semantic. The postulate of a natural property that is part of the truth-maker of disposition attributions makes it possible to explain in simple terms why the different dispositions of electric charge are indissociable: the natural property constitutes a common element of the truth-makers of attributions of these different dispositions.

As we have seen above, if an object x has a disposition D_i to manifest M_i, then x has a "causal basis" of D_i, which consists of the set B_i of all the intrinsic properties of x that contribute causally, together with the test condition S_i and background conditions, to bring about M_i. If an object has several dispositions, then each one, in general, has a different causal basis. However, when several dispositions are inseparable, in the sense that their attribution

56 Menzies (1988) offers another argument against identifying a dispositional macroproperty with the underlying microproperty. He uses an example borrowed from David Lewis, to try to show that the electrical conductivity and thermal conductivity of a metal are two dispositions based on the same set of microproperties, namely the properties of the metal's "free" electrons (i.e., electrons not chemically bound to individual atoms). Given the transitivity of identity, these dispositions cannot be identical with their common reduction base without being identical to each other, which they are not. But, on closer inspection, the reduction bases of these two dispositions are not exactly the same. In the reduction model offered by Drude in 1900, the electrical conductivity σ and the thermal conductivity κ are determined by *different* properties of the free electrons: the electrical conductivity σ is determined by the microscopic properties n (the number of free electrons per cubic centimetre), e (the unit of electric charge), τ (the relaxation time or mean free time of the free electrons: i.e., the mean interval between two collisions), and m (the mass of an electron), according to the formula $\sigma = ne^2\tau/m$ (Ashcroft and Mermin 1976, 7), whereas the thermal conductivity κ is determined by n, τ, m, and T (temperature), according to the formula $\kappa = 3n\tau k_B^2 T/2m$ (Ashcroft and Mermin 1976, 23), where k_B represents the Boltzmann constant. Block is therefore right when he notes (in correspondence with Jackson) "that cases where different dispositions seem to have the same basis, and, more generally, cases where different functional roles appear to be occupied by the same state, turn out, on examination, to involve subtly different bases and states" (Jackson 1998, 92n103). However, this case is compatible with my thesis that different properties of the microscopic parts of the metal determine different macroproperties of the metal.

to a given object always has the same truth value, we can assume that these dispositions share a common causal basis.

In the example of the dispositions associated with an electric charge, the postulate of the property of being electrically charged makes it possible to dissolve the apparent paradox of "multi-track" dispositions. The fact of possessing this property, or this power, necessarily gives the object all of the dispositions associated with it. If there are single-track dispositions, then they are a special case. However, it is unlikely that such dispositions exist. The property that would contribute to the truth-maker of the attribution of a "single-track" disposition would be a natural property not linked by laws of nature to any other properties.

The postulate of the existence of a natural property that acts as a truth-maker for attributions of disposition is fallible and subject to evaluation according to the usual criteria for the evaluation of scientific theories. There are two reasons for postulating a theoretical property: the fact that this postulate constitutes the best unifying explanation of a set of phenomena and the fact that it is fruitful for suggesting new empirical hypotheses.

One of the reasons for postulating the existence of electric charge is that it is the best unifying explanation of the fact that the above-mentioned dispositions are inseparable. The fact that this property constitutes a common truth-maker for the attribution of all these dispositions explains why every object that has one of them also has all of the others.

It is important in this context to bear in mind the distinction between the causal basis and the reduction basis. My argument provides no reason to think that the causal basis of a disposition, defined in terms of test condition and macroscopic manifestation, is microscopic.

At first glance, it might seem that multiply manifested dispositions are a superficial phenomenon that should not play any role in metaphysics. According to Bird, "we do not need to posit fundamental multi-track dispositions" (2007, 24). The reason is that it is always possible to analyze such dispositions in terms of single-track dispositions. Bird presents two reasons in favour of this thesis. The first reason is that (T1) all multi-track dispositions (which he calls "impure") are conjunctions of single-track dispositions (which he calls "pure"). According to (T1), it is equivalent to attribute to an object an impure disposition and to attribute to it a conjunction of pure dispositions. We have already developed a reason for contesting this equivalence: a theoretical power such as an electric charge is not equivalent to a conjunction of dispositions to which the charge gives rise (see Carnap 1936–37, 444–45; Mumford 1998, 41).

First, powers such as electric charge provide unifying explanations of sets of dispositions. If the possession of some electric charge were equivalent to the conjunction of the dispositions to which it gives rise, then the fact that the dispositions in this set are inseparable would be a brute fact. The postulate of an underlying power makes it possible to offer a metaphysical explanation of that fact.

Second, the idea that a power is the truth-maker of an inseparable set of dispositions accounts for the possibility of scientific discoveries. It might happen that, once the power has been postulated, it is discovered that there is in fact a larger set of inseparable dispositions of which the previously known set constitutes a subset. Let us say that electric charge was associated at some point with the dispositions (1), (2), and (3) mentioned above, before it was discovered that the set of dispositions associated with charge contained a fourth element (4). This discovery shows that it would have been wrong to consider the property of being charged as equivalent to the conjunction of the first three dispositions. It is characteristic of natural properties that it is always possible to discover new dispositions to which their possession gives rise.

The second reason for which Bird contests the fundamental status of impure dispositions is that (T2) "all impure dispositions are non-fundamental" (2007, 22), which means for Bird that they are microreducible. According to his reasoning, dispositions with multiple manifestations are reducible in terms of microscopic properties that he assumes to be necessarily "pure." In other words, they are single-track. However, as we have seen, (1) the existence of the causal basis of a disposition is independent of the issue of its microreducibility, and (2), even if a given causal basis is microreducible, there is no reason to suppose that the reduction basis is not itself a property that gives rise to many dispositions.

The reasons for postulating a natural property underlying a set of inseparable dispositions are stronger in the case of mental properties, insofar as dispositional properties such as speaking French or being shy give rise to a much larger variety of manifestations than electric charge.

My objection to Bird's thesis takes up a well-known argument against operationalism.[57] If, as is commonly the case with theoretical predicates in science, there are several operations that can serve as criteria for the application of the predicate, then one cannot identify the meaning of a theoretical predicate with an operation that determines its application conditions. It is generally acknowledged that the operationalist position is refuted by the fact that it leads to the incoherent result that the *different* operations used as criteria for the application of a theoretical concept are nevertheless, by virtue of the transitivity of their identities with the meaning of the predicate, *identical* to each other. Now the operations that serve as criteria for the application of theoretical predicates take the form of conditionals: for an object x that conducts heat and electricity, "x has a temperature of 20°C" can be justified not only by the dispositional conditional "If we put x in thermal contact with a mercury thermometer, then the thermometer will indicate 20°C," but also by the conditional "if we measure the resistance of x, then we will find $R = A \cdot 20°C$ " (where the proportionality factor A between resistance and temperature is assumed to be known). To avoid having absurdly to identify these different operations, we must consider that temperature is a property whose identity does not coincide with any of its application criteria. It is plausible that the list of criteria is open ended in general, so that the satisfaction of any one criterion is not by itself necessary or sufficient for the application of the concept. It is a good reason, but it is fallible; measurement operations can fail (or succeed for the wrong reasons). If the thermometer malfunctions, then the result of the measurement will not coincide with the real temperature, or it will coincide with it only by a fortunate coincidence.[58] Such concepts, while

---

57 See, for example, Hempel (1966). Armstrong calls Ryle's conception the "operationalist account of dispositions" (1968, 86).

58 It can also be argued (Carnap 1936–37; Stegmüller 1983, 162 ff.; Hempel 1965b) that reduction statements cannot be analytical because, when there are different reduction statements for the same disposition, their conjunction has synthetic consequences: what makes the compass needle turn also attracts iron filings. Since the reasons for being analytical are the same for all reduction statements, none is analytical. From this, we can draw the conclusion that dispositional predicates cannot be reduced analytically to observational predicates and that they designate irreducible theoretical properties. In a similar way, Mellor (1974, 175) replies to the objection of the apparent triviality of an explanation by a dispositional property that it would be valid only if that property (e.g., fragility) had as its only criterion of application the manifestation that it is supposed to explain. Dispositional properties are subject to the general criterion that is "to characterize as physically real only things that can be identified in ways other than, and independently of, the procedures

linked to conditional application criteria, cannot be equated to any one of these criteria or to their conjunction or disjunction.

Ryle is undoubtedly right to criticize "the addicts of the superstition that all true indicative sentences either describe existents or report occurrences" (1949, 108). However, we have found reasons to think that judgments that attribute a multi-track disposition can have the descriptive function that Ryle contests and that the dispositional properties thus attributed are occurrent properties. We have also seen that this puts us in a position to understand the complex and fallible link between these properties and the conditionals that serve as criteria for their attribution.

7. Conclusion

I have tried to show in this chapter that macroscopic properties such as my present intention to write the word *disposition* can be causally efficacious in bringing about their manifestations, although they can also be conceived of as dispositional properties. The defence of this thesis involves arguing in favour of the hypothesis that the categorical/dispositional distinction applies primarily to predicates and to the concepts expressed by those predicates and only indirectly to the properties designated by those predicates. If this is correct, then a dispositional predicate and a categorical predicate can designate the same property. We have seen that this conception allows us to refute a number of traditional objections against the efficacy of dispositional properties and that it allows us to escape what I have called "the epiphenomenal trilemma" with respect to macroscopic dispositional properties. According to some important and currently debated accounts, these properties are either epiphenomenal and therefore causally inert or efficacious only by being identical to microscopic properties that also constitute their reduction basis. I have shown that it is coherent and plausible to consider that a macroscopic dispositional property itself can be causally responsible for its manifestations and that the same property can be conceived of in principle in a categorical way. Whether such a property is reducible is a matter independent of the issue of its causal efficacy; however, the discovery of a reduction does not by itself

used to define those things" (Nagel 1961, 147). "If we take the, perhaps infinite, set of possible sorts of manifestation or expression of a belief that *p*, the only unifying factor we can discover in the set is that they might all spring from, be manifestations or expressions of, the one belief" (Armstrong 1973, 421).

warrant an identification of the reduced property with the reducing property. To acknowledge the causal efficacy of macroproperties does not lead to an unacceptable overdetermination of their effects. The microproperties in the reduction base cause those effects only indirectly by determining the macroscopic property. We can therefore accept the intuition that my act of writing the word *disposition* has been caused by my decision to do so. That decision is a macroscopic mental property not identical to any microscopic property of my brain.

4

Emergent Properties

1. Introduction

The aim of this chapter is to justify the causal efficacy of mental properties by presenting the hypothesis that these properties are emergent. A concept of emergence appropriate in this context must satisfy the following requirements: it must apply to certain psychological properties; it must be compatible with contemporary science; it must respect the methodological requirement not to prejudge scientific discoveries, in particular with regard to the reduction of mental properties to neurophysiological properties; and it must justify the thesis that emergent properties are causally efficacious and not merely epiphenomenal. The guiding idea is that emergence characterizes certain properties of complex objects qualitatively different from the properties of the parts of these objects and that the concept of emergence applies to properties of structured objects at all levels: the emergence of mental properties from neurophysiological properties of the brain belongs to the same category of relation as the emergence of the chemical properties of molecules from the physical properties of the atoms that make them up.

Emergent properties belong only to complex objects. There are many examples of intuitively emergent properties, including the macroscopic properties of water and ice, emergent with respect to the microscopic properties of H and O atoms; the disposition of hemoglobin molecules to bind and discharge oxygen, emergent with respect to the properties of the atomic components of these molecules; the disposition of living organisms to transmit their hereditary characteristics, emergent in relation to the properties of certain parts of their bodies and in particular in relation to the properties of DNA molecules and the genes that they contain; and, finally, the collective behaviour of a colony of ants, emergent in relation to the behaviour of the individuals that it contains (see Holland 1998; Johnson 2001).

A condition traditionally imposed on emergent properties is that of being *new*: the property E of the structured object s is emergent only if none of its parts $p_1 \ldots p_n$ has E. However, we will see that it is not easy to determine a sufficiently restrictive sense of what "new" means in this context to prevent the result that *all* properties of s count as emergent. I will propose the following criterion: a property is new if it possesses new *causal powers* that correspond to new laws.

The need to be compatible with science and not to prejudge reducibility means that we have to abandon the predominant criterion in the traditional conception of emergence: the *irreducibility* of emergent properties. Within the framework of the conception of emergence that I am going to develop, emergence is compatible with reducibility: reduction requires the discovery of a set of laws that guarantees the existence of an instance of the reduced property each time the set of reducing properties is present. This is compatible with the emergence of the reduced property: the properties of the parts of the complex object s, together with the relations existing between these parts, give rise, according to what I will call a "law of composition,"[1] to emergent properties of s. These properties are emergent in the sense that they give the whole causal powers that its parts do not have. In fact, the nomological origin of emergent properties guarantees their reducibility.

---

1 I am moving away from the terminology of Cummins (1983), which has the merit of insisting on the need to recognize non-causal forms of explanation. According to the terminology proposed by Cummins, a "law of composition" determines the analysis of a system: it is therefore the type of system, a property of the whole, that determines the type of parts. I have two objections to this interpretation. The first objection is that it invites confusion between the direction of investigation and the direction of objective determination: first we know the property of the whole; then we direct our investigations toward its components and the mode of their composition. But the concept of law requires that a law determine the objective relationships between properties, independently of epistemic priority and the direction of our investigation. If we accept that the properties of the parts determine those of the whole, then it is consistent with the ontological (i.e., realist) interpretation of laws to conceive of the laws of composition as the ontological basis of this determination: the laws of composition thus determine the whole on the basis of the parts, in a "bottom-up" direction, whereas Cummins conceives of the meaning of determination as being "top-down." The second objection is that Cummins excludes by definition the nomological analysis of multi-realizable global properties. There is no single "law of composition" (in Cummins's sense) of the property of being an eye. However, there are "laws of composition" in the sense proposed here for each type of eye.

This conception will allow us to make a qualified judgment of the possibility of *explaining* emergent properties. Reduction, always possible in principle, even if its complexity might put it practically beyond our reach, provides a scientific explanation of an emergent property. In this sense, we must reject the traditional claim that emergent properties are "inexplicable." However, it is possible to salvage the intuition that motivates that claim by applying the judgment of inexplicability, at least in part, to the laws of composition.

The new qualities of mental properties, in relation to the underlying neurophysiological properties, retain some of their intuitively mysterious character because the laws of composition that give rise to the existence of mental properties are a posteriori, and although they are necessary, like all laws (see above and Kistler 2002a, 2005a), it is *conceivable* that they are different from what they actually are. Even when we have discovered by virtue of which laws a phenomenon regularly occurs in specific circumstances, and when we have in this sense explained the phenomenon, nevertheless we can justify the intuition that we still have not *fully* explained the phenomenon, insofar as we have not explained the laws themselves that explain the phenomenon. The explanation of these laws has a limit in the axioms of the best theory,[2] which are always a posteriori and conceivably different from what they are in reality. However, the mystery of the quality of mental properties is no greater *in principle* than the mystery of the quality of chemical properties. If it is greater for us today, then it is only because we are further away from the discovery of the laws of composition that give rise to mental properties, based on neurophysiological properties, than we are to the discovery of the laws of composition that give rise to the chemical properties of molecules, based on the properties of their atomic components.

2. Minimal Conditions and Weak Emergence

The concept of emergence is supposed to do justice to a twofold conviction. The first is that emergent properties, particularly cognitive properties, are real but distinct from the neuronal properties underlying them. According to the second, they are determined exclusively by the underlying material

2 According to the "best system account" of the laws of nature, due to Mill and Ramsey and more recently developed by David Lewis, "a contingent generalization is a *law of nature* if and only if it appears as a theorem (or axiom) in each of the true deductive systems that achieves a best combination of simplicity and strength" (1973, 73).

properties; in particular, cognitive properties are determined by the properties of the brain of the subject as well as those of the rest of her body and the environment with which she interacts. This conviction amounts to giving cognitive properties an intermediate status: they are distinct from the properties belonging to the parts of the subject, in particular the brain and the parts of the brain, as well as from those properties of the subject of the same quality as the properties belonging to its parts, such as her mass or volume. However, they are not *independent of* her brain and, more generally, of her body and environment. To say that the perceptual state of a person looking at a red tomato is an emergent property is to say that only a cognitive system obeying precise architectural constraints can possess it. Yet nothing simpler can have this experience; in particular, none of the parts of the system can have it. For example, the visual system alone, isolated from the rest of the brain and body, cannot have such experiences. This is what we mean when we say that the perceptual state is not a "resultant" property: a property of an object is resultant if it is not qualitatively different from the properties of its parts. Conversely, however, the perceptual state is determined exclusively by physical and physiological conditions of the brain, alongside the rest of the body and its particular environment. In this sense, emergentism is opposed to dualism: an emergent property is different from the properties of the substrate that brings it into existence, but it is this substrate and its physical environment that exclusively determine its nature.

The task of characterizing emergence can therefore be broken down into two parts. The first and easier part consists of distinguishing the emergent properties of an object from hypothetical "dualist" properties, whose existence and nature would be determined by something other than the material properties of the object's parts and their interactions. We can establish this distinction by means of three constraints that a property must satisfy in order to be emergent. In other words, the existence of a property that violates one of these constraints can be accommodated only within a dualistic framework.[3] Conversely, the fact that a property satisfies these constraints is not enough to make it emergent in the strict sense: these are necessary but not sufficient conditions. They are also satisfied by resulting properties and in particular, in the case of properties of people, by their purely physical and physiological

3 The satisfaction of these constraints corresponds to what Stephan calls "the weak theory of emergence" (1999, 66–67).

properties. The more difficult task will be to distinguish emergent properties from these resultant properties. The three constraints are as follows.

First, an "anti-dualist" condition that emergent properties must satisfy is that they belong only to physical systems (i.e., systems composed exclusively of physical parts). Everything that is emergent — whether property, process, or structure — belongs to a system made up exclusively of physical parts. This condition makes emergence compatible with physicalism.

Second, a property *P* of a complex system can be called "emergent" only if it does not also belong to parts of the system. Having a certain mass is a property that belongs to a person as a whole, but it also belongs to its parts. This is enough to justify the intuition that the property of having a mass does not "emerge" at the level of the cognitive system, because it is already present at the lower levels of complexity to which its parts belong. We can express this condition by saying that emergent properties are *systemic* properties of an object, which means that no part of the object possesses them (Stephan 2006).

Emergent properties are "anti-homeomerous." To take Armstrong's definition, "a property is homeomerous if and only if for all particulars, *x*, which have that property, then for all parts *y* of *x*, *y* also has that property" (1978, 2: 68). Emergent properties are anti-homeomerous in the sense that a necessary condition of emergence is that, for all *x* that possess the property, *none* of *x*'s parts *y* possesses it.

Third, the two previous constraints are not strong enough: if a cognitive system possessed a "spiritual" property imposed on it by God, either directly (as is the case according to occasionalism) or indirectly (as is the case according to parallelism), rather than through the material substrate of the system and its organization, then it would not be "emergent" from this material substrate. It would not be emergent even if God imposed the property in question only on systems with physical parts (so that the property satisfies the first condition) and if he imposed it only on the whole system but not on any of its parts (so that it satisfies the second condition). An emergent property must be determined exclusively by the parts making up the system: an emergent property is *determined synchronously* by the properties of the parts of the system and their organization. This determination requires *mereological supervenience.*[4] The emergent properties of a system *supervene* on the properties of its

4 See Kim (1984, 165; 1988b). I return to the concept of mereological supervenience in Chapter 5.

parts: there can be no change in the supervening properties without a change in the properties of the parts of the system.[5]

It is important to distinguish the *synchronic* determination of an emergent property from its *causal* determination. The metaphysical analysis of causality is controversial. Insofar as we conceive of cause and effect as particular events that occupy a delimited portion of space-time, cause and effect must be spatiotemporally distinct.[6] When we speak of a *property F* that causes another *G*, *F* must be the property of a cause event *c*, and *G* must be the property of an effect event *e*, where *c* and *e* are spatiotemporally distinct. In terms of causality between events, it is excluded conceptually that the cause is the same event as the effect; it is also excluded that cause and effect partially overlap. Causality can be said to exist only if the cause is spatiotemporally located elsewhere than the effect. On the basis of these conceptual constraints on causality, the emergence of *A* from *B* is incompatible with the existence of a *causal* dependence of *A* on *B*. According to the third condition, emergence is a synchronic relation. *A* is an emergent property of system *s* if *A* is determined by the properties *B* that the system possesses *at the same time*. Furthermore, *A* belongs to the whole system *s*, and the properties *B* belong to the parts p_i of the same system: the parts p_i possess *B* at the same time as the system *s* possesses *A*. Accordingly, there is partial spatiotemporal overlap between the carriers p_i of the properties *B* and the carrier *s* of the emergent property *A*; the emergence of *A* from *B* is a form of non-causal determination.

In such a conceptual framework, it is difficult to interpret Searle's assertion that "consciousness is a causally emergent property of systems" (1992,

5 However, synchronic determination goes beyond mereological supervenience: the latter is compatible with the possibility that systemic properties are determined by God rather than by the properties of the parts of the system (Kim 1993a). I will develop this point later in this chapter.

6 I cannot justify this thesis here. See Kistler (1999b, 2006d). Hume (1978, 76) justifies it as follows: if simultaneous causality were possible, then there could be no non-simultaneous causality. A cause that is sufficient for its effect could not precede it by a finite time because, since it is sufficient, no additional condition intervenes at the moment when the effect occurs. But in this case there is no sufficient reason for the effect to occur just then and not earlier or later. A different and less direct way of refuting the possibility of simultaneous causality is to argue that causes and effects are events: that is, particular entities that occupy particular areas of space-time. In this framework, simultaneous causality is impossible because the effect can occupy neither the same space-time zone as the effect nor a different one: the first case is impossible because the effect must be distinct from the cause; the second is impossible because the physics of relativity forbids simultaneous action at a distance.

112). Searle holds that all mental states and processes are at least potentially conscious, which leads him to defend more generally the thesis of the "causal emergence" of all mental phenomena. They are, he says, "caused by neurobiological processes" (1992, 1, 115, and passim).

Searle's thesis undoubtedly has its source in the traditional but erroneous identification of nomological determination with causal determination. Not every determination according to laws of nature is causal (see Humphreys 1989, 300–01; Salmon 1990, 46–50; Kistler 1999b, 2004a, 2006d). Only the neglect of this distinction can explain why Searle moves without justification from the claim that emergent properties are determined by causal interactions between the parts of a system to the claim that this determination itself is causal: emergent properties, he says, "have to be explained in terms of the causal interactions among the elements" of the system that possesses them, which he takes to justify calling them "causally emergent system features" (1992, 111). However, the fact that emergent properties are *synchronous* with the properties of the parts of the system that determine them forces us to conclude that this determination follows non-causal laws of determination. Emergent properties, therefore, are *not caused* by the properties that determine them through laws of composition.

In the same vein, E.J. Lowe justifies his assertion that mental properties have causal powers "not wholly grounded in . . . the causal powers of those elements of the system which produced [them]" (1993, 636–37) by comparing consciousness to a spider's web. Unlike the liquidity and transparency of water, causal powers based entirely on the powers of water's constituent molecules, the spider's web has "a life of its own" (636), an expression that Lowe borrows from Searle, who says that "consciousness gets squirted out by the behavior of the neurons in the brain, but once it has been squirted out, it then has a life of its own" (1992, 112). This analogy clearly reveals the confusion shared by Searle and Lowe between causal determination and non-causal nomological determination.[7] The spider's organs produce the web in a process spread out over time: it begins with events that take place in the spider and ends in the existence of the web. This means that the events that cause the web take place earlier than the events in which the web exercises its causal powers, for example, when it supports the spider passing across it.

7 I develop this criticism of Lowe's conception of mental properties in Kistler (2005b, 2022).

Along these lines, the analogy between the web and mental properties, in their respective relationships with the spider and the brain, is misleading: the subject possesses a mental property by virtue of a synchronic determination, which gives mental properties particular powers, but no independent existence in a substantial sense. Conversely, once created, the web no longer depends on the spider; it can continue to exist even if the spider disappears, whereas cognitive properties are permanently dependent on the underlying brain processes and events that give rise to their existence. As soon as brain activity ceases, the mental property ceases to exist. Synchronous nomological determination is not peculiar to the relationship between brain and mind. This is how, for example, the properties and processes taking place at the level of the electrons in a metal determine, in a non-causal way, the electrical and thermal conductivity of the metal. Without the microscopic processes involving the electrons, there is no electrical conductivity. This is also a case of non-causal determination, because the metal has the conductivity and the underlying microscopic properties at the same time.[8]

According to the concept of emergence defended here, emergence does not exclude reduction. However, an emergent property can only be reduced to its basis provided that we do not conceive of reduction as the discovery of the identity of the reduced property with its reduction basis. Insofar as one subscribes to the thesis[9] that reduction leads to the discovery of an identity, the compatibility of emergence with reduction can lead to strange conclusions. Searle, presupposing his thesis that emergence is a causal process, concludes that, once their reduction is complete, emergent properties "can . . . be identified with their causes" (1992, 115). This conclusion destroys the intelligibility of both emergence and causation. They would be reflexive relations, where things can cause themselves and emerge from themselves. It is possible to avoid such absurd consequences if we accept that neither the reduction of *A* to *B* nor the causality between *A* and *B* entails the identity of *A* and *B* and if we conceive of emergence as a relation of non-causal determination of the systemic properties of a complex object from the properties of its parts.

8 If we accept the idea that a "temporal part" of an object is an event, then these are properties of the same event. This appears to be a natural consequence of the conception according to which the content of a well-defined spatiotemporal zone always constitutes an event, even if it corresponds to no apparent change. See Kistler (1999b, 2006d).

9 It is defended in particular by Causey (1977). See Chapter 1.

3. Broad and the Epistemic Conception of Emergence

Properties that obey the three conditions formulated above can be said to be emergent in a weak sense. These are properties that belong only to exclusively physical objects, and they are determined exclusively by the properties of their parts. The weakness of this first concept of emergence lies in the constraint of "novelty." For example, without further clarification of what counts as "new" or "qualitatively different" in relation to the properties possessed by the parts of an object with emergent properties, the distinction between the "resultant" property of a human being of weighing 70 kg and her "emergent" property of hoping that the war will end soon has no rigorous basis. The general property of being massive is certainly not systemic because all of the parts of the body also possess it. However, weighing 70 kg is formally a systemic property because no proper part of the person's body possesses it. We have the intuition that this property is not "qualitatively different" or "qualitatively new": this intuition justifies the introduction of the distinction between strongly emergent properties that are not only systemic but also qualitatively new in relation to the properties of the parts and only weakly emergent properties, such as the property of weighing 70 kg.

The British emergentists at the end of the nineteenth century and the beginning of the twentieth century tried to make this distinction rigorous by analyzing it in terms of *explanation*: "resulting" properties are those whose presence can be explained on the basis of the properties of the parts, whereas an analogous explanation of the presence of emergent properties is possible only with the help of ad hoc postulates. However, this way of grounding the distinction between emergent and resultant properties is incompatible with physicalism. According to physicalism, every emergent property conforms to the third condition mentioned above: it is determined exclusively by the properties of the parts of its possessor. It is therefore sufficient to discover the laws underlying this nomological determination to explain (in the sense of deductive-nomological explanation) the presence of the emergent property. Within the framework of physicalism, all global properties can be explained in principle. Accordingly, this possibility does not give rise to any relevant distinction among such properties, between emergent and resultant.[10]

10 We cannot hope to construct a relevant ontological concept from the epistemic criterion according to which irreducible predicates and laws are emergent. However, the epistemic conception

The relevant laws that I call — in the words of the last great British emergentist C.D. Broad — "laws of composition" determine a property of a complex object on the basis of the properties of its parts and the interactions between these parts. Once the law of composition that determines a given emergent property is known, that property, and the laws in which it is involved, can be explained and reduced.

The physicalist framework — and in particular the third condition of the nomological determination of the properties of complex objects — thus imposes the possibility of emergent properties being subject in principle to reductive explanation. However, this constraint seems to clash with a central thesis of the emergentist tradition, according to which the distinguishing mark of emergent properties is the *impossibility* of explaining them completely. Broad (1925, 65) resolves the tension between the constraints of nomological determination and the impossibility of explanation by conceiving of emergent properties as nomological, in the sense that their presence is determined systematically by a law of nature. The presence of the emergent property can be predicted and even explained, therefore, at least in the minimal sense of deductive-nomological explanation. What cannot be explained in the case of an emergent property is the law that determines it itself. This law remains a "raw nomological fact."[11]

Broad gives the example of silver chloride. It is on the basis of the properties of its components, chlorine and silver, and their relationship as components of a molecule that has no other components, that a law determines the properties of silver chloride. But this law is an experimental law that is "brute" (Broad 1925, 55) in the sense that it cannot be derived within a more general theoretical framework. It can be discovered only experimentally by observing the properties of samples of this compound. Insofar as it is a purely experimental law in this sense, it is inexplicable. There are no answers to the questions why is silver chloride composed of equal parts of chlorine and silver, and why does it have such or such properties? The law of composition transfers its inexplicable, and in this sense mysterious, character to the

---

of emergence proposed by Hempel and Oppenheim (1948) is still widely accepted. Churchland expresses it by saying that "claims about the emergence of certain properties are therefore claims about the relative poverty in the resources of certain aspirant theories" (1985, 12). See also McLaughlin (1992). Malaterre (2010) shows that such an epistemic conception of emergence can be illuminating in analyzing the scope and limits of scientific explanation.

11 I take this expression from McLaughlin (1992, 81).

emergent properties whose existence it determines. As we will see, this thesis is responsible for the marginalization of emergentism in the philosophy of science. It can already be found in J.S. Mill, for example in his assertions that "it is impossible to deduce all chemical and physiological truths from the laws or properties of simple substances or elementary agents" and that "the laws of Life will never be deducible from the mere laws of the ingredients" (1843, Book III, Chapter VI, para. 2). Broad expresses it in these terms: "The characteristic behaviour of the whole *could* not, even in theory, be deduced from the most complete knowledge of the behaviour of its components" (1925, 59).[12] The properties of the compound AgCl, for example, emerge, according to this criterion, from the properties of its atomic components (Ag and Cl), insofar as the law of composition of AgCl is "a law which could have been discovered only by studying samples of silver-chloride itself, and which can be extended inductively only to other samples of the same substance" (Broad 1925, 65).

Broad proposes to ground the distinction between emergent and non-emergent properties on the distinction between two types of a posteriori laws, the criterion being their integration into a theory. Broad contrasts principles of composition (1925, 45, 66), which are explanatory, with "trans-ordinal laws" (77 ff.), which are "brute nomological facts" because they cannot in turn be explained on the basis of more general laws. Broad knows that the principles or laws of composition[13] are a posteriori, just as much as trans-ordinal laws, and that both types of law determine the properties of the whole entirely from the properties of the parts. According to Broad, the difference between the two types of law lies in the possibility of deducing them in turn

12 The rest of the quoted sentence contains a crucial qualification that makes it possible to reconcile emergentism with contemporary science and to which I will return. Broad anticipates my thesis that it is not possible to deduce the emergent properties of a complex object from complete knowledge of the properties of its parts, insofar as they are the properties that the parts possess *in isolation* or in other combinations. I say "anticipates" because he expresses this thesis in the language of causal determination, whereas it is a form of non-causal determination.

13 Broad seems to think that all laws of composition have an additive form but without explicitly affirming it (1925, 66). McLaughlin notes the existence of this "gap in the discussion" (1992, 77) but does not appreciate its importance. McLaughlin does not note that quantum mechanics refutes the emergentist thesis only if it is assumed that the laws used to deduce molecular properties from atomic properties are "compositional principles." Since these principles are not additive, the original conception leads us to consider this deduction as trans-ordinal. It is only because of the imprecision of what counts as a "compositional principle" in Broad that McLaughlin can conclude that the quantum deduction of the molecular state uses only compositional principles and thus succeeds in refuting the thesis that chemistry is emergent.

from more general laws. A trans-ordinal law allows predictions to be made and practical control to be exercised over the emergent properties that it determines. In these respects, it is no different from a law of composition. The distinctive property is that the trans-ordinal law that determines the global properties of a given sort of compound object is a primitive law that applies only to that type of object. It can be discovered only experimentally by examining samples of that particular compound (Broad 1925, 65).

This way of distinguishing between emergent and non-emergent properties, based on the distinction between primitive experimental laws and deducible composition laws, makes the concept of emergence an *epistemic* concept relative to a theory — in other words, relative to a given moment in the history of science. A property *E* is emergent only in relation to a given theory *T*. To say that *E* is emergent, relative to *T*, simply means that *T* does not have the conceptual resources to derive the law of composition that gives rise to *E*. As soon as a more powerful theory is constructed that has the means to produce this derivation, *E* ceases to be emergent. In other words, according to the conception of emergence developed by Mill and Broad, it seems to be reasonable to expect that no property is emergent in an absolute or ontological sense. For any property that appears to be emergent relative to the theories available at a given moment, it is only a matter of time before a theory is discovered that can explain the laws that give rise to its existence. Emergence is an effect of perspective: emergent properties are those whose scientific explanation is still unknown at a given moment.[14]

14 The epistemic conception of emergence was proposed by Henle (1942) and by Grelling in his correspondence with Hempel and Oppenheim. We could interpret Lewes's remark that we will "some day, perhaps, be able to express the unseen process [by which hydrogen and oxygen are transformed into water] in a mathematical formula; till then we must regard the water as an emergent" (1875, 414; cited in Stephan 1992, 28) as the expression of an epistemic concept of emergence. According to McLaughlin, Lewes does not maintain that the emergent character of a property depends on the progress of our knowledge. The passage quoted only expresses the idea, compatible with an ontological interpretation of emergence, that the evaluation of the hypothesis that a property is emergent depends on the progress of our knowledge. See Stephan (1992, 28n4). Similarly, Broad says that "within the physical realm it always remains logically possible that the appearance of emergent laws is due to our imperfect knowledge of microscopic structure or to our mathematical incompetence" (1925, 81). Broad does not express a relativist doctrine of emergence like that of Hempel and Oppenheim (1948) but merely the recognition that our hypotheses about the emergent or resultant nature of a given property are fallible, even though their truth conditions are objective and absolute.

In an epistemic or relativist conception of emergence, it could certainly be argued that, at a given moment, there are always laws that have the status of brute experimental laws, whose only justification comes from induction on the basis of observation or experimentation. For a trans-ordinal law that indicates the properties of a whole as a function of the properties of its parts, the former properties count as emergent but only in relation to this stage of scientific progress or in relation to the limited knowledge that characterizes it. According to the research strategy of seeking to integrate knowledge of isolated experimental laws within increasingly general theories, we can expect the law eventually to change its status in relation to Broad's classification. As soon as there is a theory that allows it to be deduced from more general principles, the experimental law *becomes* a law of composition. The properties of silver chloride that Broad (1925, 64) gives as an example are emergent only in relation to the scientific knowledge of 1925. Progress in chemistry has made it possible to deduce a certain number of properties of this compound from general laws that apply to all halogen and metal compounds and then from even more general laws that determine the properties of a compound from the structure of the electronic orbitals of its constituent atoms.

This observation does not settle the question, of course, of whether there are experimental laws that will definitively resist integration into a broader theory. But Broad's distinction seems to be capable of characterizing only properties emergent relative to a given level of scientific knowledge. Insofar as the strategy of scientific research is dominated by the "mechanistic" paradigm in Broad's sense, which seeks to deduce any law first discovered experimentally within a theory from more general laws, we can never infer from the "emergence" relative to a given stage of scientific knowledge to objective or ontological emergence, independent of the level of knowledge.[15]

Let us return once more to the crucial point, the possibility in principle of deducing the existence of a given property. The British emergentists are materialists who explicitly take a stand against dualist positions. For example, Broad explicitly rejects the doctrine of "substantial vitalism" defended by Driesch, who postulates the existence of entelechies to explain the phenomena

15 According to Hüttemann and Terzidis (2000), the properties that are emergent according to Broad's criterion are "anomalies." Insofar as scientists pursue a mechanistic strategy, they set out to forge a theory that transforms trans-ordinal laws into laws of composition and consequently deprives emergent properties of this status.

of life (see Broad 1925, 58, 69; McLaughlin 1992, 86). Therefore, phenomena that occur at levels of complexity higher than that of atomic physics are not independent of phenomena that occur at lower levels. To postulate substances specific to each level would be tantamount to considering that these phenomena are independent. This is exactly what Broad criticizes as "Theories of a Special Component" (1925, 60; see also 55–58), which explain the behaviour of complex objects on the basis of "the presence of a peculiar component which does not occur in anything that does not behave in this way" (55; see Beckermann 1992, 16). Science does not support the postulate of such substantial components, such as the *élan vital* or the entelechies of the vitalists. In this sense, classical emergentists accept the idea that complex phenomena are determined nomologically by the properties of lower levels:[16] trans-ordinal laws are laws. On the ontological level, they determine emergent properties; on the epistemological level, knowledge of a trans-ordinal law allows the deduction of the presence of an emergent property from the observation of the presence of the conditions that appear in the law's antecedent. Insofar as the Hamiltonian of a system composed of two hydrogen atoms determines nomologically the formation and properties of the molecular ion H_2^+, it is possible to deduce (using other laws and principles of quantum mechanics) the formation of the molecular ion H_2^+, even if this deduction involves a "trans-ordinal" law (because the form of the Hamiltonian has not yet been deduced from more general principles).[17]

There seems to be a contradiction, in the doctrine of Broad, between the postulate of the existence of trans-ordinal laws — which determine the existence of emergent properties — and the assertion that it is impossible, even in principle, to deduce the existence of these properties or to predict the situations in which they are exemplified. As Broad puts it,

> if the emergent theory of chemical compounds be true, a mathematical archangel, gifted with the further power of perceiving the microscopic structure of atoms as easily as we can perceive hay-stacks, could no more predict the behaviour of

16 Provided that we accept the thesis of the necessity of the laws that determine emergent and resultant properties, emergentism therefore leads to local strong emergence. I will come back to this later in this chapter.

17 I return to this example later in this chapter.

silver or of chlorine or the properties of silver-chloride . . . than we can at present. And he could no more deduce the rest of the properties of a chemical element or compound from a selection of its properties than we can. (1925, 71)

But the contradiction is only apparent. The solution lies in the sentence omitted from the quotation above: it is not possible to deduce the properties of silver chloride "without having observed samples of those substances" (71). In other words, emergent properties can indeed be predicted and explained but only by means of trans-ordinal laws that are irreducible experimental laws. These laws are "brute nomological facts" that themselves cannot be explained.

We have already seen one reason why this criterion is not adequate to the ontological concept of emergence that we are seeking. The progressive construction of theories regularly leads to the integration of experimental laws into theories, which makes it possible to transform their status of "brute nomological facts" into theorems deducible from axioms or principles of a more general order.[18] A second reason is that there are usually several equivalent formulations of the same theory, which attribute the status of axioms to different statements.

McLaughlin (1992) proposes to interpret Broad's criterion ontologically: trans-ordinal laws are objectively and absolutely primitive in the sense that in principle they cannot be derived within a more general theory. This ontological interpretation runs the risk of rendering the concept of emergence empirically empty, whereby it would remain as a coherent concept but void of any positive reason to think that it has application.[19] According to McLaughlin's interpretation, the emergentist doctrine bases the distinction between compositional principles and trans-ordinal laws on the distinction between particle pair forces and *configurational* forces. According to emergentists, each level of organization is characterized by specific properties that are causal powers and obey laws specific to their level of organization. According to some emergentists, in particular Sperry, these causal powers and the laws that characterize them are based on "configurational forces"

18 This is why Hempel and Oppenheim take Broad's thesis that emergent properties are those determined by a trans-ordinal law — that is, a law that cannot be deduced but must be induced directly from experience — to be "untenable" (1948, 262).

19 This is also the conclusion of Hüttemann and Terzidis (2000).

(1964; cited in Sperry 1986, 266):[20] that is, "*fundamental* forces that can be exerted only by certain types of configurations of particles, and not by any types of pairs of particles" (McLaughlin 1992, 52).[21] Now the conception of the distinction between compositional principles and trans-ordinal laws, which identifies the former with laws governing forces between pairs of particles and the latter with laws governing configurational forces, amounts to a refutation of emergentism by stipulation. For it is then enough to note that contemporary science, and above all quantum mechanics, give us every reason to believe that there are no configurational forces, to deduce that there are no emergent properties (McLaughlin 1992, 89–91).

Nagel (1952), Feigl (1958, 411–13), and McLaughlin (1992) note that contemporary science — notably quantum mechanics and the theory of evolution by natural selection — has come to account for and explain some of the properties that the emergentists Mill and Broad considered to be paradigms of emergent properties. They draw the conclusion that the success of these scientific theories removes all credibility from the hypothesis of the existence of absolutely emergent properties. This argument effectively refutes the versions of emergentism defended by Bain, Lewes, Alexander, Lloyd Morgan, and Broad as well as those defended later by Popper and Sperry. However, we will see that it is still possible to salvage the intuition that, in complex systems, new properties appear or "emerge" with causal powers different from the powers of the parts of these systems. In other words, it is possible to construct an ontological concept of emergence that takes account of scientific progress.

To explore the possibility of a concept of emergence compatible with contemporary science, we will take a closer look at an example that is crucial both systematically and historically: the case of chemical bonding. At a systematic level, the determination of the chemical properties of molecules from the physical properties of the atoms that make them up is a paradigm of the relationship of determination between adjacent levels of compositional complexity. We can then examine the hypothesis according to which the determination of biological properties from the chemical properties of the

20 On Sperry, see Stephan (1992, 43n20).

21 McLaughlin attributes this interpretation to classical emergentism and to Broad in particular: "Broad's Emergentism commits him to configurational forces" (1992, 88). It does not seem to me that the text justifies this interpretation. McLaughlin does not give a textual justification, as he is careful to do for his other assertions. But this point of exegesis is of secondary importance.

components of living bodies, and the determination of psychological properties from biological properties, fall on the same side of the great division between resultant and emergent properties as the determination of the chemical properties of molecules from the physical properties of their atomic components. In historical terms, it was the advent of quantum mechanics, and in particular the explanation of the origin of chemical bonding on the basis of physical laws, that tolled the bell for the great era of British emergentism. In McLaughlin's words, it was "no coincidence that the last major work in the British Emergentist tradition coincided with the advent of quantum mechanics. Quantum mechanics and the various scientific advances it made possible are arguably what led to British Emergentism's fall" (1992, 54). The quantum-mechanical explanation of elementary chemical properties, therefore, can serve as a test case for concepts of emergence. An acceptable conception of emergence must be compatible with contemporary science in general and with quantum mechanics in particular. If such a conception exists, then the verdict of Feigl, Nagel, and McLaughlin is premature: only certain historical versions of emergentism are obsolete, notably Broad's, but not emergentism *simpliciter.*

4. Strong Emergence in Terms of the Impossibility of Deduction

Insofar as all properties of complex systems are determined by laws of composition, it seems to be promising to try to ground the distinction between emergent and resultant properties in the nature of laws of composition. For the moment, we can rule out one interpretation of what a law of composition is. According to a thesis of logical empiricism, also at the origin of an important interpretation of reduction, they are *metalinguistic* principles whose role is to *define one predicate* in terms of others.[22] This doctrine is rooted in

22 I have analyzed this conception of reduction — considered by Nagel (1961, 354–57) — in Chapter 1. Wimsatt (1976a, 221) explains that the *many-many* relationship (see also Endicott 1998) between micro- and macroproperties in the reduction of Mendelian biology to molecular biology (see also Hull 1974) refutes the conception of reduction as "translation." Armstrong's (1968) and Kim's (1998) theories of functional reduction share the conclusion that mental and neurophysiological predicates are simply two sets of predicates that designate the same properties. The dissimilarity is that Armstrong and Kim take the difference between these two sorts of predicates to be a logical difference: second-order predicates contain a generalization over first-order predicates.

a more general conception of science as a set of statements; according to it, laws are theorems deducible from axioms or fundamental principles, which are also statements. Nomological statements include causal laws, which are a posteriori and express factual regularities. Laws of composition, in opposition, are taken to be (conventional) rules of translation. In an extensionalist interpretation of the language of science, a universal law of biconditional form expresses not the co-extensionality of *two* properties but the extensional equivalence of two predicates designating *a unique* property. In this sense, laws of composition are taken to be metalinguistic statements.[23] In this metalinguistic conception of reduction, a property is emergent if there is no rule for translating the predicate that designates it into a predicate constructed exclusively from terms of a reductive science. Hempel and Oppenheim (1948) point out that this doctrine renders theses concerning the emergence of a property trivial. The fact that we have not yet discovered the laws of composition that determine certain biological or psychological properties is merely a sign of the terminological incompleteness of present-day science. In their words, "in this interpretation, the emergent character of biological and psychological phenomena becomes trivial; for the description of various biological phenomena requires terms which are not contained in the vocabulary of present-day physics and chemistry" (263).

There are more general reasons against applying a metalinguistic interpretation to laws of composition. Let me briefly mention some reasons for adopting a realist conception of properties, according to which laws are constraints that these properties exert on each other. Predicates designate these properties, and nomological statements designate the laws (see Armstrong 1983; Kistler 1999b, 2006d). Nomological statements are a posteriori even though they designate necessary relations (Kistler 2002a). Such statements, therefore, are neither conventional nor metalinguistic, both in the case of causal laws and in the case of compositional laws. Laws of composition, in particular, *determine* the properties (emergent or resultant) of a complex object from the structural properties of its components. However, the existence of such a determination relation does not justify the idea that the properties

23 The linguistic interpretation of the distinction between emergent and non-emergent properties can be found in Pap (1951–52), Tully (1981), and Teller (1992). In Tully's words, "if secondary quality terms are indefinable in terms of microscopic particles, then this is a logical fact about the language we use to describe the world" (266; cited in Stephan 1992, 40).

of the system are identical to the properties of the components. The non-identity of the properties renders inadequate the conception according to which the statement that expresses their nomological correlation is metalinguistic, in the sense of expressing the definition of one predicate by another, where the two predicates designate the same property.

With this in mind, let us return to the traditional conception of emergence in terms of the impossibility of deduction. I will evaluate this conception by examining the question of whether hydrogen's property of forming stable molecules is an emergent property, using the following criterion:

> (C1) *Criterion of deducibility*: a global property *G* of a complex object *s* is *(strongly) emergent* if and only if (it is weakly emergent,[24] and) it is impossible in principle to deduce (i.e., explain, predict) the fact that *s* possesses *G*, from *complete information about* the components of *s* and *the properties possessed by the parts of s when they are isolated.*

We will see that, in the light of contemporary science, the applicability of the concept of emergence on the basis of the criterion of deducibility depends crucially on the interpretation of the expression "complete information."

The advent of quantum mechanics seemed to refute directly the thesis that the chemical properties of molecules are emergent according to (C1), with respect to the physical properties of the atoms that make them up: in quantum mechanics, certain simple but fundamental properties of molecules can be deduced from general principles and certain premises concerning the component atoms.

It might come as a surprise to read about the debate on the emergence of chemical properties from physical properties in an inquiry into the status of psychological properties in relation to neurophysiological properties. However, this issue is of fundamental importance to the latter problem insofar as the hypothesis according to which the mental emerges from the physiological will gain credibility by virtue of the justification of a more general hypothesis. According to this general hypothesis, emergence makes it possible to characterize the nature of the relationship between the properties of

24 Earlier I defined the weak emergence of a property *E* of a system *s* in terms of three conditions: *E* belongs only to objects composed solely of physical parts; *E* is systemic; *E* is determined entirely by compositional laws, based on the parts of *s* and their interactions.

a domain of objects x in relation to the properties of the objects of which the objects x are composed. Conversely, if this general hypothesis proves to be untenable in the fundamental case of the relationship between physical and chemical properties, then the hypothesis according to which psychological properties are emergent in relation to the properties of the neurons in the brain is likely to appear ad hoc.

To evaluate the thesis that quantum mechanics refutes emergentism with regard to chemical properties (according to the criterion of deducibility), I will examine the most fundamental and simple deduction of all explanations of chemical properties, based on physics: the deduction of the stability of the molecule H_2 or, more precisely, of the molecular ion H_2^+. We will then see that the viability of the emergentist thesis depends on the interpretation of what counts as "complete" knowledge of the properties of the components. If we understand knowledge of the parts of the complex object H_2^+ — the H atoms — to be absolutely complete, then this knowledge contains knowledge of the law according to which these atoms form, in precise circumstances, H_2 molecules. The global properties of these molecules, therefore, will not be emergent according to the criterion of deducibility. If, however, we consider that knowledge of the parts is complete as soon as it contains knowledge of all the laws that determine the evolution of the parts — the H atoms — *in isolation*, then even quantum mechanics does not allow us, on the basis of such limited knowledge, to deduce the formation and properties of the H_2 molecule. In that case, the criterion of deducibility leads to the result that the global properties of molecules are emergent after all.

It therefore appears that the anti-emergent consensus in contemporary analytical philosophy is built upon a strong interpretation of *completeness* in the definition of the distinction between emergent and resultant properties. If we consider that complete knowledge of the properties of components is not limited to the properties that these components possess in isolation, but extends to the laws that govern their interactions, then the refutation of the thesis that the stability of the H_2 molecule is an emergent property becomes trivial: if knowledge of the law of formation of H_2 is already part of the "complete" knowledge of H atoms, then by stipulation we have transformed the property of stability of H_2 into a property deducible from physics.[25] I will

25 Hempel and Oppenheim (1948, 260) credit Kurt Grelling for first noting this point in the course of their correspondence. They point out that Grelling was murdered by the Nazis before

therefore adopt a more restrictive interpretation of the completeness of knowledge of the properties of parts.

Quantum mechanics allows us to deduce that molecules exist from the existence of atoms and their properties. The paradigmatic deduction concerns the prediction of the existence, or more precisely the permanent existence or stability, of the hydrogen molecule H_2. The limits of the possibilities of exact calculation are quickly reached, but the conceptual importance lies in the possibility, in principle, of deducing a property that belongs to the chemical level of the organization of matter, from knowledge of objects belonging to a lower level of complexity (e.g., atoms). According to this fundamental result, quantum mechanics makes it possible to deduce that two hydrogen atoms, whose nuclei are close enough to allow a partial superposition of the space occupied by their respective electrons, form a stable H_2 molecule. It is advantageous to study the nature of this deduction in the case of the molecular ion H_2^+, the simplest system of the molecular level existing in nature: whereas the molecule H_2 has two electrons and two atomic nuclei, each consisting of a proton, the molecular ion H_2^+ has only one electron for two nuclei. It is therefore by studying the explanation of the existence of this molecular ion H_2^+ that we can hope to isolate the features of a deductive explanation that crosses the boundary between different levels of complexity.

The deduction of the existence of the hydrogen molecular ion is based on three nomological presuppositions. The first presupposition concerns the structure of the system's Hamiltonian. The Hamiltonian is an operator that determines, via the Schrödinger equation, the energy levels of the system. Its form is characteristic of the system under study. The system of the molecular ion H_2^+ consists of two protons and one electron. Its Hamiltonian $\mathcal{H}$ contains a term corresponding to the kinetic energy of the electron ($p^2/2m$) (where p represents the momentum and m the mass of the electron) and three terms corresponding to electrostatic interactions among the three objects making up the system: the repulsive interaction between the two protons, separated from each other by R (e^2/R, where e represents the reduced charge,

---

he could publish these ideas (245n1). Including all second-order nomological properties in what is required to know a property perfectly is less radical, however, than including knowledge of all properties of everything. As Van Cleve points out, we could trivialize the doctrine of emergence even further: "I could maintain that all properties of everything in the universe are deducible from the properties of James Van Cleve, provided you counted among my properties such items as 'being such that the Eiffel Tower is 1,056 feet tall'" (1990, 223).

with $e^2 = {q^2}/{4\pi\varepsilon_0}$), and the attractive interaction between the electron and each of the two protons from which it is separated by r_1 (e^2/r_1) and r_2 (e^2/r_2) respectively.[26]

$$(*)\ \mathcal{H} = p^2/2m - e^2/r_1 - e^2/r_2 + e^2/R$$

The second presupposition concerns the fundamental law of quantum mechanics, according to which the energy levels of a system are the solutions of Schrödinger's equation

$$\mathcal{H}\,\psi = E\psi\,,$$

where $\mathcal{H}$ denotes the Hamiltonian, E the energy, and ψ the wave function, which characterizes the state of the system.

The third presupposition concerns the general applicability of the quantum mechanical formalism to solve this equation in the case of the system under consideration.

The law that determines the Hamiltonian according to equation (*) is a law of composition. It determines a characteristic property of the system as a whole, as a function of certain properties of its components and their relationships. The properties of the components involved, on which the properties of the whole depend exclusively, are the electric charges of the electron and the protons and the mass of the electron, m.[27] The relationships that determine the state of the overall system are the respective distances among these three components: R, r_1, and r_2. The law of composition (*) is empirical or a posteriori, determining the impact on the energy of the overall system of the individual properties of the components and especially their mutual relationships. The law (*) determines the structure and properties of the system as a function of the interactions among its components, themselves

26 The equation (*) should be interpreted as describing quantum operators.

27 The mass of the protons is not involved since the movement of the protons is neglected. The mass of the protons is much greater than that of the electron, which means that the electron's motion is much faster than that of the protons. This is "why, to a first approximation, the two motions can be studied separately" (Cohen-Tannoudji, Diu, and Laloë 1977, 1: 511). This is the so-called Born-Oppenheimer approximation (see also Cohen-Tannoudji, Diu, and Laloë 1977, 2: 1160).

determined by the properties of the individual components and their mutual relationships (in this case, spatial distances).

In contrast to classical emergentism, I consider the possibility that the properties of complex objects are determined by non-causal laws. Here we need to pay attention to a distinction that I mentioned earlier: the formation of a molecule from two originally isolated atoms is a causal process that evolves through time. We can consider that the two H atoms gradually come closer together until their electron orbitals partially overlap. But this causal process is not determined by the law (*) that determines the energy levels of the system as a function of the distance *R* between the nuclei and as a function of the Hamiltonian of the system, determined in turn by laws governing the interactions among the properties of the parts of the system. The determination of the system's energy levels and the distance R_0 at which this energy is minimal is not a process that evolves through time, in which case it would make it possible to justify the distinction between temporally separate cause and effect. The calculation proceeds in stages and takes time, but as Duhem (1906, 25) points out this does not warrant concluding that there is a real process whose stages correspond to the stages of the calculation. The calculation takes into account the electrical charges of the parts of the system as well as the fact that, when the electron orbitals partially overlap, a quantum "resonance" phenomenon occurs that leads to a decrease in the energy of the state of the system, thanks to the possibility that the electron occupies a hybrid state around the two protons.

To give just the general idea of this calculation, it is assumed that the state ψ of the electron must be an eigenstate of the Hamiltonian $\mathcal{H}$.[28] In the method of variations, it is assumed that this state ψ results from a linear superposition of the states ψ_1 and ψ_2, the stable states of an electron bound to one of the H atoms in the absence of the other. We know that the ground energy state is a minimum of the mean value of $\mathcal{H}$, where that minimum is a function of the distance *R* between the two protons, considered as a parameter. This minimum corresponds to the eigenvalue *E* of the energy of the system as a function of *R*. It turns out that the function $E(R)$ has a minimum. This can be explained by the presence in the energy term of a "resonance integral," which represents a lowering of the energy due to "the possibility

28 For details, see Cohen-Tannoudji, Diu, and Laloë (1977, 1: 409–10; 2: 1159–71).

for the electron to 'jump' from the vicinity of one of the protons to the other" (Cohen-Tannoudji, Diu, and Laloë 1977, 2: 1167). The existence of a value of R for which E is minimal corresponds to the existence of a stable ψ state resulting from a linear superposition of the two states ψ_1 and ψ_2, which correspond to the localization of the electron around one of the two protons. The stability of this superposed state, in turn, explains the existence of the molecule: the movement of the two protons away from each other, as well as toward each other, moves the system away from its state of minimum energy.

The stability of the molecular ion H_2^+, the distance R_0 (which characterizes the stable state), and the shape of the molecular electron orbital are deducible. But it is crucial to evaluate the information contained in these premises so as to discern whether this means that these global properties are emergent according to the criterion of deducibility. If that information bears only on the properties that the components possess when they are isolated, then the deduced global properties are not emergent; however, if the possibility of deduction requires that the premises contain information about the laws that determine the global properties from the interactions among the components, then these global properties are emergent according to criterion (C1) (although they are deducible).

The form of the deduction of the existence of a stable state in which the two H atoms are linked into a molecule corresponds to the second situation: according to (C1), the stability of the H_2 molecule is an emergent property. The possibility of deducing the stability of the molecule requires knowledge not only of the properties of the isolated atoms but also of the *laws of interaction* among the components of *different* H *atoms*. Insofar as the stability of the molecule cannot be deduced from the properties that its components possess *in isolation*, it is emergent.[29] Furthermore, the laws of interaction cannot be deduced from the laws determining the evolution of the components in isolation.

However, there are reasons to think that the criterion (C1), which made it possible to obtain this result, is inadequate. It appears that the condition expressed by (C1) is too strong when we reconsider the examples considered

29 Grelling and Oppenheim (1937–38, 1939) draw a parallel between the emergent properties of complex objects and the phenomena studied in *Gestaltpsychology*. The parallel lies in the fact that the prediction of a Gestalt as well as that of an emergent property "requires knowledge of certain structural relations among its parts" (Hempel and Oppenheim 1948, 261n18) and cannot be obtained from knowledge of the isolated parts alone.

above: (C1) makes all of the properties of complex objects emergent because knowledge of the isolated evolution of the parts never allows us to deduce the consequences of their interactions. The form of the laws of interactions is never an a priori consequence of the laws of the isolated evolution of the parts. As Broad says, "it is clear that in *no* case could the behaviour of a whole composed of certain constituents be predicted *merely* from a knowledge of the properties of these constituents, taken separately, and of their proportions and arrangements in the particular complex under consideration" (1925, 63). Predicting the result of an interaction always presupposes knowledge of an a posteriori law. Predicting the behaviour of the whole presupposes "that we have found a general law connecting the behaviour of these wholes with that which their constituents would show in isolation"; this law is the "law of composition" specific to the system (63).

Criterion (C1) must therefore be rejected. In fact, it does not offer the means to ground the intuitive distinction between emergent and resultant properties because it imposes such a strong condition on non-emergence that all global properties end up appearing as emergent.

Given this observation, Broad suggested an alternative criterion that takes account of the fact that deducing a global property of a system necessarily requires information about the laws governing the interaction of the parts. According to Broad's criterion,

> (C2) the global property *G* of a complex system *s* is (strongly) emergent if and only if (it is weakly emergent and if) the law of composition the knowledge of which makes it possible to deduce the presence of the property *G* is a *law specific to the type of system s which cannot be derived from laws applying to parts of s in isolation and from laws of composition specific to other types of system.*

(C2) differs from (C1) only by the reference to knowledge of the behaviour of the parts *in other combinations*. Consequently, the properties of AgCl would not be emergent according to (C2) if they could be deduced from compositional laws for *other* molecules containing Cl and Ag. However, it is doubtful that this corresponds to a real relaxation of the conditions of non-emergence. The possibility of using knowledge of the behaviour of Cl and Ag in other combinations, for example in sodium chloride NaCl (common salt), to deduce the behaviour of silver chloride necessarily requires knowledge of general laws that apply — in this example — to all molecules of which chlorine is a component. Furthermore, knowledge of the laws of composition concerning a set of types of molecules including the Cl component (NaCl, HCl, etc. but not AgCl) does not logically imply either the law of composition of compounds outside this set, such as AgCl, or general laws that apply to all Cl compounds. This is the problem of induction. The knowledge of the laws of composition of several types of molecules including Cl provides only the premises for an *inductive* argument leading to such a general law. Accordingly, it appears that the criterion of non-emergence (C2) proposed by Broad — despite the introduction of the clause "or in other combinations" — is just as strong as criterion (C1). Furthermore, (C2) is too weak because according to it all properties of complex systems come out as emergent, just as in (C1).

However, my reasoning indicates a criterion that is stronger than (C1) and (C2): we can hypothesize that the non-emergent properties are those that can be deduced from the complete information on the components of the system in isolation, as well as from *general laws* determining the interactions of these components, without recourse to laws specific to the type of system in question.

> (C3) A global property *G* of a complex system *s* is (strongly) emergent if and only if (it is weakly emergent and if) the law of composition — the knowledge of which makes it possible to deduce the presence of the property *G* — is a law specific to the type of system *s that cannot be derived from general laws applying to s as well as to other types of systems.*

> (C3 — non-emergence) A global property *G* of a complex system *s* is non-emergent if and only if the law of composition — the knowledge of which makes it possible to deduce the presence of the property *G* — can be derived from general laws applying to *s* as well as to other types of systems, without it being necessary to call on a law specific to the type of system *s*.

According to this criterion, which expresses a necessary condition for emergence, the property *G* is emergent only if the law that gives rise to *G* is a brute experimental law: it cannot be derived from more general laws (i.e., laws that apply to the properties of parts), even outside the type of system that possesses

G. On the contrary, a property of a complex object determined by *general laws* is not emergent. If the law of composition can be derived from general laws that apply to the properties of parts even outside the particular system under consideration, then the property is explicable to the same extent as the law. It is therefore non-emergent according to criterion (C3).

This new criterion of non-emergence is really weaker than criteria (C1) and (C2). According to (C3), the properties of the molecular ion H_2^+ are no longer emergent with respect to the properties of its components because their derivation only calls on general laws determining the components of the Hamiltonian of any system possessing components of the same types. However, (C3) still seems to be too strong as a criterion of emergence (or too weak as a criterion of non-emergence): it is true that many global properties of complex systems, including chemical properties, cannot be explicitly derived at present from knowledge of the parts and from general laws. But at least as far as chemical and biological properties are concerned, the successes of deductive explanations of certain paradigmatic properties — such as the stability of the H_2 molecule and the ability to transmit hereditary traits — constitute paradigms around which research programs aimed at reductive explanations in these fields have been built. From the point of view of such a research program, the systems that count as emergent according to (C3) appear to be emergent only provisionally or epistemically. (C3) seems to analyze a *concept of provisional or epistemic emergence*. From this point of view, properties that cannot (yet) be deduced appear as "anomalies" (Hüttemann and Terzidis 2000, 274) bound to disappear as soon as general laws and adequate deductions are discovered. From the point of view of the research strategy of reductionism that bets on the existence of such laws and deductions, it seems to be reasonable to presume that in the long run the chemical properties of molecules will prove not to be absolutely emergent according to (C3), compared with the physical properties of the atoms that make them up: today's quantum mechanics allows us to deduce certain simple but fundamental properties of molecules from general principles and premises concerning the component atoms. Other chemical properties seem to differ only in the complexity of their derivation, not by any fundamental ontological difference. Ontologically, (C3) is too weak a criterion for non-emergence because all complex properties satisfy it; in other words, it is too strong a criterion for emergence because no complex property satisfies it.

Another way of seeing that (C3) is too strong a criterion for emergence is this. We are looking for a criterion to justify the intuitive distinction between emergent properties that are qualitatively new and only appear in systems satisfying precise structural constraints and resulting properties that belong to complex systems but are only quantitatively different from properties belonging to their parts or to simpler systems. This distinction is ontological, not epistemic. Belonging to one or another of these categories must therefore be independent of our theories. We can be wrong, of course, about whether a property is emergent or resultant. However, our intuitive concept of emergence is incompatible with a conception that makes the emergent or resulting character of a property systematically relative to the scientific theories accepted at a given moment in history. Yet this is a consequence of (C3). Let us say that property *G* is determined by the law of composition *C*. *C* can be more or less integrated into a theoretical system. It might start out as a "brute" empirical postulate and later become derivable from other empirical laws, which in turn are brute regularities without explanation, before a way of deducing *C* from the most general axioms and principles of theory is finally discovered. As *C* is deductively integrated into an increasingly powerful body of theory, *G* gradually loses its emergent character and is transformed into a resultant property.

We must therefore find a criterion that does not categorize all macroscopic properties as emergent — as in the cases in (C1) and (C2) — but that also does not categorize them all as resultant — as in the case of (C3).

5. Emergence as Non-Aggregativity

William Wimsatt (1986) suggested a fruitful way of conceiving of systemic properties that, though explicable and therefore in a sense reducible to the properties of the system's components and their interactions, are nevertheless emergent, in the sense of appearing only in systems with a specific composition and organization. Wimsatt identified four features that distinguish the properties that he called "aggregative"[30] from emergent properties.

The *emergent* properties of a complex object depend on the organization of its parts. A property of a complex object is *aggregative* if it does not depend

30 The term "aggregative" is equivalent to the term "resultant," used to designate non-emergent properties in the tradition of British emergentism.

on the organization of its parts. The mass of a pile of stones is a paradigmatic example of an aggregative property. This mass does not depend on the spatial arrangement or interaction among the stones that make up the pile. The heap can be dismantled and reassembled at random without any change in the overall property of its mass.

Of course, we could have excluded the mass of the heap of stones from the emergent properties for the simple reason that it is not systemic: the mass of the pile is a property *of the same kind* as the mass of the individual stones. However, the application of this criterion remains intuitive until we have a clear criterion of what counts as "the same kind" of property. The criteria proposed by Wimsatt are intended to remedy this lack of clarity: according to his first criterion, a global property not qualitatively different from the properties of its components is recognized by the fact that its presence does not depend on the organization of its parts and therefore does not undergo any modification when these parts are swapped. The mass of a heap of stones is aggregative because it does not vary when two stones exchange their positions. The capacity of a given portion of DNA, conversely, is a non-aggregative property according to this criterion: the expression of a gene depends on its location downstream from a control sequence. Accordingly, substituting one sequence for another can modify the conditions of its expression. However, according to this criterion, non-aggregativity is not necessary for emergence. Wimsatt identifies four ways in which a property of a complex system can be emergent or non-aggregative. Each is sufficient, but none is necessary, for a property to be emergent. Invariance with respect to the permutation of parts is not necessary for emergence because there are emergent properties — such as the transparency or the rhombohedral shape of a quartz crystal — not modified by the permutation of parts. Permuting molecules or parts of the crystal does not affect these overall properties.

According to Wimsatt's second criterion, a property is aggregative if it changes quantitatively but not qualitatively when parts of the system are added or removed. The stability of an arch composed of stones with a trapezoidal cross-section is non-aggregative according to this criterion, insofar as the arch loses its stability if one of the stones in it is removed. However, this criterion is not a necessary condition for emergence either. The transparency and rhombohedral shape of a quartz crystal are emergent properties in the sense that the microscopic components of the crystal have neither transparency nor rhombohedral shape nor any qualitatively similar property. But the crystal remains transparent and rhombohedral even if some molecules are removed or added.

There is a third way in which a property can be non-aggregative: it is possible for an overall property to be modified without adding or subtracting parts (as in the second criterion) and without permuting components (as in the first criterion). It can be modified by changing the spatial organization of the parts. According to this criterion, the ability to lead to the expression of a phenotypical trait is a non-aggregative property of genes, insofar as a change in *the spatial arrangement* of genes — for example, because of recombination — affects the expression of a gene depending on the presence of the appropriate control units at precise locations upstream from the DNA sequence. This criterion is a generalization of the first. It does not provide a necessary condition for emergence for the same reason as the first; some emergent properties resist decomposition and rearrangement of parts.

According to Wimsatt's fourth criterion, a global property of a system is non-aggregative if its existence depends on interactions among the parts of the system. The interaction among the four subunits of a hemoglobin molecule, for example, reduces the energy required to bind oxygen; this capacity is therefore a non-aggregative, or an emergent, property according to this criterion. All emergent properties are undoubtedly dependent on such interactions. However, not all interactions give rise to emergent properties: the mass of the heap of stones remains an aggregative property of the heap even if the stones attract each other according to gravitation. This criterion is therefore necessary, but not sufficient, for emergence. Therefore, we need to add a condition to specify which interactions are sufficient for emergence. This is the purpose of the criterion that I consider in the next section.

6. Emergence in Terms of Non-Linear Interaction and Mill's Principle of the Composition of Causes

If we accept the idea that all global properties are determined by general laws from the properties of the components, then it seems to be promising to try to ground the distinction between emergent and resultant properties on the mathematical form of these general laws of interaction. We can draw inspiration from a fundamental distinction introduced at the origin of the emergentist tradition by John Stuart Mill. He distinguishes two ways in which two (or more) interacting causes can determine their common effect. In the mechanical mode of interactive determination, the effect is the mathematical sum — arithmetic or vector — of the effects that each cause would have had if it had acted alone. Mill characterizes effects determined in this way by saying that they "obey the principle of the Composition of Causes" (1843, Book III, Chapter VI, para. 2). He distinguishes them from effects determined according to a "second mode" that corresponds to "a breach of the principle of Composition of Causes" (1843, Book III, Chapter VI, para. 2). Using the terminology introduced by Lewes (1875), emergentists later called effects determined in the first additive way "resultant effects." If the effect of several causes is not the mathematical sum of the effects that would have resulted from the separate actions of the causes, Mill says, then the determination of the effect obeys a "heteropathic law" (Book III, Chapter VI, para. 2) and produces a "heteropathic effect" (Book III, Chapter X, para. 4). In the vocabulary of Lewes, we may speak of "emergent law" and "emergent effect."

We can use this distinction between two types of determination as a basis for the distinction between emergent and resultant properties. However, this presupposes that we dissociate the Millian distinction from the analysis of *causal* laws and apply it to compositional laws.

The law of composition (itself reducible to the laws determining the interactions among the parts of a complex object or system) expresses the dependence of the global properties of the complex object on the properties that the parts possess in isolation. The distinction between emergent global properties and resulting global properties can be grounded on the mathematical form of the interaction laws as well as the law of composition. If the interaction laws have an additive or linear form, then the law of composition will also be linear. According to the criterion inspired by Mill, the properties determined by a linear law of composition are resultant, and the properties determined by non-linear laws of composition are emergent (Wimsatt 1996, S374).

> (C4) A global property G of the complex object s is (strongly) emergent if and only if (it is weakly emergent and if) the fact that s has G is determined by the fact that s has the parts $p_1 \dots p_n$ having the properties $P_{11} \dots P_{nm}$ as well as by a law of composition that is not a logical consequence of the laws governing the properties $P_{11} \dots P_{nm}$ of the parts in isolation. The non-linearity of the law of composition is itself a consequence

of the non-linearity of the laws of interaction applying to the properties of the parts.

Mathematically, the general form that characterizes different forms of addition is the linear form $y = ax+b$. The total force acting on a body is a paradigmatic resultant property. It is determined by a linear superposition of the component forces. This criterion, therefore, justifies the intuition that the total force acting on a body *results from* the parallel action of the component forces. (C4) also justifies the intuition that the stability of the hydrogen ion H_2^+ is emergent with respect to the properties of the atomic level:[31] the Hamiltonian that determines the state of the molecule does not result from the addition of the Hamiltonians determining the state of each of the component atoms. I will come back to this in a moment.

For a systemic property to be qualitatively different from the properties of the parts of the system, the law of composition determining it must have a non-linear form. In Holland's words, "emergence is above all a product of coupled, context-dependent interactions. Technically these interactions, and the resulting system, are *nonlinear.* The behavior of the overall system *cannot* be obtained by *summing* the behaviors of its constituent parts" (1998, 121–22). The "context" on which emergent properties depend might lie within the system: in this sense, the context in which the parts are located is constituted by the other parts with which that part interacts. It can also be the "extra-systemic" context with which certain parts of the system interact. However, the second case can be reduced to the first case by broadening the contours of the system so that it includes what was originally considered to be extra-systemic.

Criterion (C4) makes emergence compatible with physicalism. All properties of complex objects are determined by empirical laws. The difference between resultant and emergent properties does not concern the possibility

31 Curiously, McLaughlin uses two different criteria for emergence at two different points in the same article, to arrive at opposite results, without noting the contradiction. When he sticks to the criterion established by Mill, Lewes, Alexander, and Morgan, according to which the existence of a property is a heteropathic effect of lower-level properties (i.e., the property is emergent) when it is not determined by a law of additive composition, McLaughlin notes that "the Emergentists were right about there being emergents" (1992, 75); however, when he uses the criterion that he attributes to Broad, according to which a property is emergent if its causal power obeys a law of "configurational force," he arrives at the result that there are no emergent properties in this sense since there are no configurational forces (89–91).

of explaining and predicting them.[32] Provided that the laws that determine them are discovered, all of them can be explained and predicted in principle. The difference lies only in the form of the laws. Laws of composition that have the form of a linear function give rise to resultant properties; laws of composition that have a non-linear form give rise to emergent properties. The emergent properties of complex objects are qualitatively different from the properties of their parts because they are determined by non-linear laws of composition, but they are not irreducible.[33] By making emergence compatible with the possibility of reducing even emergent properties, and thus with physicalism, (C4) expresses a weaker concept than the traditional one that requires inexplicability, irreducibility, or unpredictability.

(C4) has retained Mill's distinction between linear and non-linear laws. However, there are some important differences between (C4) and Mill's original criterion. First, as we have seen, (C4) distinguishes between resultant and emergent properties as a function of the mathematical form of *non-causal* compositional laws, whereas Mill's distinction concerns causal laws and forms of *causal* determination. Second, Mill confuses the *ontological* distinction between laws of different forms (the distinction between linear and non-linear laws) with an *epistemic* distinction.

Let us consider the first difference. Mill distinguishes between two forms of causal determination: the difference is "between the case in which

---

32 Prediction and explanation are not always equivalent. Within the framework of the deductive-nomological model (and if we are dealing only with deterministic laws), explanation and prediction are analyzed in the same terms of the logical relationship (that of a valid deduction) between premises and conclusion. Only the direction in which knowledge is acquired differs. In explanation, we already know the conclusion, and we learn from which premises it can be deduced. In prediction, we learn which conclusions follow from premises already known. In the deterministic context, it is therefore legitimate to consider, as Broad does (see Stephan 1992, 38), the concepts of prediction and deduction (or deductive explanation) as equivalent. However, when the law is probabilistic (or statistical), it might be possible to explain an event because it can be deduced that it would happen with a certain probability. The question of whether this probability should be greater than a certain universal value is controversial (see Salmon 1990). Conversely, insofar as the probability is not 1, the event nevertheless can be considered unpredictable. It is conceivable that probabilistic composition laws exist (see Stephan 1992, 33). But this does not seem to be the case for the laws that interest us here. In general, therefore, it is not necessary to insist on the difference between prediction and explanation. This difference can be taken to be epistemic or pragmatic, in the sense indicated above.

33 Chalmers (1996, 378n41) and Bedau (1997, 375) call concepts of emergence of this kind "innocent."

the joint effect of causes is the sum of their separate effects, and the case in which it is heterogeneous to them" (1843, Book III, Chapter VI, para. 2). The former mode of composition of causes obeys the principle of the composition of causes, whereas the latter mode violates this principle. In contrast, (C4) uses the criterion of the mathematical form of laws to distinguish two forms of non-causal determination. The properties of complex objects are determined by the properties of their components and their interactions, according to non-causal laws of composition, specific to those components and interactions. Let us therefore ignore the fact that for Mill and other emergentists nomological determination is identified with causal determination, to reinterpret the terminology of Lewes (1875) in a non-causal way: properties determined by a linear law of composition are "resultant," whereas properties determined by non-linear laws are "emergent."

It is only because (C4) transposes the Millian distinction to non-causal laws that we can use it to ask whether the determination of the Hamiltonian of the molecular ion H_2^+ from the properties of its components obeys the "additive mode"[34] of (non-causal) determination or the "heteropathic mode." Stripped of its causal interpretation, Mill's question becomes that of whether the determination of the molecular properties of H_2^+ is obtained by means of an addition — vectorial or algebraic — or by means of a more complex law of composition. The form of

$$(*)\ \mathcal{H} = P^2/2m - e^2/r_1 - e^2/r_2 + e^2/R$$

shows that the interactions between the components of the molecule obey not an additive law but a "heteropathic" law. The Hamiltonian of H_2^+ is not the sum of the Hamiltonians of each separate body. For this reason, the properties of the molecule determined by this law are "emergent" rather than "resultant."

Second, criterion (C4) differs from Mill's in that it is *ontological*, whereas Mill gives his criterion an epistemic meaning. For him, the distinction

34 It is impossible to use the term "mechanical" in our conception of emergence. Being essentially a form of causal determination, a non-causal mechanical determination would be a contradiction in terms. This is not the case with the term "chemical," only contingently associated with a mode of causal determination.

between linear and non-linear laws is equivalent to the distinction between global properties that can be deduced a priori from knowledge of the parts in isolation and global properties whose existence can be discovered only a posteriori, through experience.[35] In his words, "it is impossible to deduce all chemical and physiological truths from the laws or properties of simple substances or elementary agents" insofar as the laws "of chemistry and physiology . . . owe their existence to a breach of the principle of Composition of Causes" (1843, Book III, Chapter VI, para. 2). In particular, "the Laws of Life will never be deducible from the mere laws of the ingredients" (1843, Book III, Chapter VI, para. 2). The thesis of the impossibility in principle to deduce chemical and physiological laws from physics makes Mill's conception incompatible with physicalism.

However, Mill is wrong to regard the ontological distinction between two forms of law as equivalent to the epistemic distinction between two forms of explanation. In reality, no law of interaction is a logical consequence, knowable a priori, of the set of laws governing the evolution of properties in isolation (apart from interactions). In other words, if we take as resultant only those properties *G* of *s* determined *logically* from the properties of the components of *s*, then all properties would be emergent. Without any empirical law, no property of *s* can be deduced. Indeed, even the additive composition of properties that gives rise to a resultant property is not a priori: it merely obeys a particularly simple law.[36] The mode of additive composition is not distinguished by its a priori character: we cannot know a priori how two bags of flour of 1 kg mass each will interact when the grocer places them together on the scale, any more than we can know a priori how the photons of superposed

35 Popper (1977) seems to confuse these two distinctions (between a priori and a posteriori and between linear and non-linear) insofar as he considers as emergent all properties *P* of complex objects *x* not logically (or a priori) deducible from the properties of the components of *x*. Conversely, Broad acknowledges that these distinctions are not equivalent. I will return to this issue in a moment.

36 It is perhaps this point that Stephan has in mind when he says that, "to deduce the weight of the whole, one must also invoke the principle of the additivity of weight. That principle, while nomological, is logically contingent" (1992, 35). According to the conception of laws that I have defended elsewhere (Kistler 2002a, 2005a), laws are metaphysically necessary. The difference between logical principles and laws of nature is therefore epistemic. We can know the former a priori but the latter only a posteriori.

rays of light will interact.[37] Knowledge of the laws of interaction is always a posteriori.[38]

Emergent properties appear if a non-linear law governs the interaction.[39] For example, the mass of a solid object is smaller than the sum of the masses of the atoms that compose it: the stability of a solid object originates from the fact that, in the solid state, the energy of the system is less than the sum of the energy of the atoms. This reduction in energy is equivalent to a small reduction in mass. Therefore, the mass — and consequently the weight — of a macroscopic body is emergent in relation to the masses of its atomic components.

Kronz and Tiehen (2002) criticize the idea of grounding the concept of emergence of a property *G* on the non-linearity of the laws of interaction that generate *G* by an analysis of the origin of the existence of entangled states in quantum systems, which are paradigmatic cases of emergent states. According to them, "the mark of a non-emergent property of composite systems in quantum mechanics crucially involves a *multiplicative* operation,

---

37 Nagel expresses this point as follows:

> When the matter is viewed abstractly, the "sum" of a given set of elements is simply an element that is *uniquely determined* by some *function* (in the mathematical sense) of the given set. . . . [T]he question [of] whether such a function is to be introduced into a given domain of inquiry, and if so what special form is to be assigned to it, cannot be settled a priori. (1952, 23)

Nagel uses the example of the luminosity of a surface illuminated by two sources: "The physical brightness of a surface illuminated by two sources of light is sometimes said to have for one of its parts the brightness associated with one of the sources" (19). He then remarks (22) that one can speak sensibly of the sum of the two luminosities only if the light is monochromatic.

38 Broad (1925, 62 ff.) explicitly recognizes that principles of composition are logically contingent and a posteriori. Therefore, he does not explain the greater transparency of resulting properties compared with emergent properties, which corresponds to the possibility of explaining the former, but not the latter, by the particular (namely, a priori) epistemological status of principles of composition. Hempel and Oppenheim's (1948) argument — taken up by Van Cleve (1990, 224) — that Broad is wrong to consider the mass of a compound object as a resultant because the law of the composition of masses is contingent is therefore based on a misunderstanding (McLaughlin 1992n38). Hempel and Oppenheim wrongly think that Broad takes the principles of composition of resulting properties to be necessary. I can add that Hempel and Oppenheim and Van Cleve are wrong in their judgment of the modal status of the law of composition. Van Cleve argues that "the parallelogram law for the composition of forces is logically contingent" (224), concluding that the Broadian conception of emergence makes the behaviour of a three-body system emergent. The law of composition of forces is an empirical law of nature. If we substitute "a posteriori" for "logically contingent," then we obtain a correct judgment in line with Broad's doctrine.

39 Non-linearity is necessary for emergence. I do not claim that it is sufficient.

factorability into tensor product vectors (in the case of states) or matrices (in the case of properties), rather than an *additive* one" (2002, 333; italics added). Indeed, the mathematical operation that represents the transition from the description of two separate systems to the description of a complex system is the tensor product. If F_1 and F_2 are vectors, each of which represents a two-state system, then the vector representing the complex system formed from these two systems is $F_1 \otimes F_2$.

The entangled states of complex quantum systems, which, according to Kronz and Tiehen (2002) correspond to emergent properties, are distinguished by the impossibility of expressing them as tensor products of lower-dimensional states. This observation warns us against an abusive simplification of the criterion of emergence. It refutes the idea that any state of a complex system, whose description can be obtained from a multiplicative procedure based on the description of its constituent parts, is emergent. However, the objection raised by Kronz and Tiehen against such a simplistic conception does not call into question my criterion of emergence, which concerns the form of the law of composition and itself is a consequence of the set of laws that governs the interactions among the parts. In quantum mechanics, this law is represented by the Hamiltonian of the system, which determines its evolution and properties, in particular the eigenvalues of its observables. And it is indeed the presence of non-linear terms in the Hamiltonian that is responsible for its non-separability, which in turn is responsible for the non-separability of the system and the fact that the state of the intricate system is emergent. This means that we cannot represent the system and its evolution as a conjunction (whether by addition or by multiplication) of subsystems, each of which evolves as a function of its own Hamiltonian, which is "separable" (i.e., whose expression is independent of references to the other subsystems). The properties of the parts of such an intricate system cannot be represented without reference to the other parts. In quantum mechanics, a part p_1 of a system *s* is described by a "density operator" $\rho_1(t)$ obtained by forming the trace over the complementary part p_2 of the system. The mathematical development shows that the evolution of p_1 can be described by a separate operator of evolution $U_1(t)$, only if the Hamiltonian is separable. If it is not separable because it contains non-linear terms of interaction, then the evolution of p_1 is at all times dependent on the evolution of the whole system *s*. Mathematically, the operator U(t) of the evolution of the whole system is separable: that is, it can be represented as the product of two operators $U_1(t)$

and $U_2(t)$ so that $U(t) = U_1(t) \otimes U_2(t)$, if and only if the Hamiltonian of *s* is separable: that is, it can be written as the sum of two Hamiltonians $\mathcal{H}_1$ and $\mathcal{H}_2$ describing separately the subsystems p_1 and p_2. In mathematical terms, we switch from addition to multiplication because U is an exponential function[40] of $\mathcal{H}$: $U(t) = \exp(-i\mathcal{H}t)$.

Given the fundamental role of the Hamiltonian, this seems to show that the criterion of the additivity of laws of interaction is applicable in quantum mechanics. However, the criterion can be generalized outside that framework.[41]

7. Qualitative and Quantitative Difference

The qualitative "novelty" (Alexander 1920, 14n2) of the emergent properties of a whole in relation to the properties of its parts is one of the most important criteria of emergence for the classical emergentists. Novelty is also the fundamental feature of emergence in the definitions of emergence offered by Bunge (2003, 17)[42] and Sperry (1986, 267).[43]

In line with my thesis, Bunge argues that the emergence of a property is compatible with the possibility of explaining or deducing it: "It is mistaken to define an emergent property as a feature of a whole that cannot be explained in terms of the properties of its parts. Emergence is often intriguing but not mysterious: explained emergence is still emergence" (2003, 21).[44] However, the concept of novelty used by Bunge is too vague to support the distinction between emergent and resultant properties. According to him, a property *P* of a complex object *x* is resultant if there are components of *x* that also have

40 This presupposes that $\mathcal{H}$ does not depend explicitly on time.

41 Kronz and Tiehen give "some measure of plausibility" to the idea that the presence of non-linear terms in the Hamiltonian of a classical system can be responsible for emergent properties insofar as "a classical system can exhibit chaotic behavior only if its Hamiltonian is nonseparable" (2002, 332).

42 Bunge does not distinguish, in this definition, between synchronic and diachronic emergence. His conception of emergence makes any property acquired over time, even during a simple spatial shift, an emergent property.

43 In Popper's words, "there is the emergence of life . . . [that] creates something that is utterly new in the universe" (1977, 342). A little later, he says that "the fact of the emergence of novelty, and of creativity, can hardly be denied" (343).

44 Bunge also says that "every emergent property of a system can be explained in terms of properties of its components and of the couplings amongst these" (1977, 503; quoted by Stephan 1992, 31).

the property *P*. In other words, Bunge identifies emergence with what I called above the systemic character of a property. But we have seen that being systemic is not enough to be emergent. Without a distinction between systemic and emergent, Bunge cannot avoid considering all macroscopic properties as emergent. The property of weighing 100 kg is "new" compared with the property of weighing 50 kg, whereas it is clearly a resultant property. For his criterion to acquire a precise content, it is necessary to give a rigorous meaning to the distinction between qualitative novelty and quantitative novelty so that it corresponds to the intuitive difference between emergent and resultant properties.

Assumption (C4) suggests that the non-linear form of the law of composition is responsible for the qualitative difference between the emergent properties and the properties of the parts. Psychological research has established the existence of psychophysical laws that can be described via functions from stimulus space to representational space. As we saw earlier (Chapter 3.5), the metric properties of the latter can be determined empirically (Shepard 1962; Clark 1993). With knowledge of these psychophysical laws of composition, it becomes possible to predict the secondary quality or the qualitative appearance associated with a stimulus that has not yet been experienced (by some particular subject or even by anyone), for example the smell of a perfume made for the first time. As Feigl says, with knowledge of psychophysical laws and of the representational space of odours, "we should be able to predict the location of the quality in the topological space of odors" (1958, 416) by means of extrapolation.

The phenomena of colour vision can illustrate the appearance of new properties: the qualities of experience or "qualia." When we see light produced by the superposition of rays of different colours, the perceived colour is qualitatively different from the colour of the components. It is not the result of a simple superposition, or addition, of the sensations produced by each of the stimuli taken separately. The phenomenal appearance of the perceived colour is the result of a complex process. The signals received by the light-sensitive cells in the retina — the cones — are transformed by a set of "opponent processes," postulated in psychological terms by Hering (1920). The neurophysiological mechanism underlying this process was discovered by Hurvich and Jameson (1957). Through the interplay of reinforcement and inhibition, the three types of light-sensitive cones located in the retina, with maximum sensitivity respectively to long (*L*), medium (*M*), and short (*S*) wavelengths, give rise to three signals: an achromatic transmission channel made up of neurons that "add" the signals from the *L* and *M* receptors, giving rise to the perception of white when the signal is positive, and the perception of black when the signal is negative. A second, chromatic, channel is made up of neurons that "subtract" the signals from the *L* and *M* receptors. The signal from this channel produces perceptions of red, if the net signal is positive, or perceptions of green, if the net signal is negative. Subtracting the signal from the receptors sensitive to the shortest wavelengths (*S*) from the sum of the signals from *L* and *M* gives rise to a third signal that produces perceptions of yellow, if the result of this interaction is positive, or blue, if it is negative (see Lennie 2000, 577–78).

This mechanism explains why it is impossible to perceive a colour that is a mixture of red and green or of yellow and blue: when we are exposed to light composed of a superposition of rays of two exactly opposite colours (in the sense that their signals come from the same place in the visual field but are of opposite signs) and of the same intensity, the signal is zero. The opposite colours, rather than producing a mixed or combined appearance, cancel each other out. As Hardin puts it,

> since the neutral point of an opponent pair is achromatic, any stimulus that will put both chromatic opponent systems in balance will yield an achromatic perception. . . . If we are given a yellowish green light, we can render the appearance of this stimulus achromatic by adding enough blue to offset the yellow and enough red to balance the green. . . . Whiteness may be generated by as few as two wavelengths and in an indefinitely large number of ways. (1988, 38–9)

It is equally interesting to disregard the details of the complex mechanisms that lead from the absorption of light of different wavelengths to the perception of colours and instead to examine the result of this process: the structure of colour representations. As we have seen in Chapter 3, it is possible to reconstruct the structure of the cognitive space of colour representation, and in particular the minimum number of its dimensions, just from judgments of similarity between perceived colours. We find that the colour representation space has two dimensions and that the colour representations of the visible colours occupy points on the circumference of an approximately circular structure (see Figure 3.1).

The net result of the treatment of the physical stimulus by the colour vision system, therefore, is the transformation of a one-dimensional space into a two-dimensional structure located in a two-dimensional space. The simplest physical stimuli that give rise to the perception of colours are "monochromatic" rays in the physical sense (i.e., rays that contain only one wavelength). These coloured stimuli are ordered in a single dimension, which corresponds to a portion of the series of wavelengths, starting at a wavelength of around 400 nm and ending at a wavelength of around 700 nm.[45] The net result of the treatment of light signals by the visual system is the emergence of coloured perceptions.

The psychophysical laws that determine colour representations from physical stimuli are clearly non-linear since they associate representations located around the circumference of a circle with stimuli ordered along a single dimension. More generally, Shepard's work on the structure of the representational space of a certain number of phenomena belonging to different sensory modalities, shows that the transformation between stimuli and representations does not preserve relations of proximity. Intuitively, the projection of the space in which the stimuli are located onto the representational space is accompanied by a deformation that is generally inhomogeneous: the representations of two pairs of stimuli at equal distance from each other in physical space generally will not be at equal distance from each other in representational space.

However, the non-linear form of the law of composition is not a sufficient condition for qualitative novelty. According to the Weber-Fechner law, the intensity of sensations is a logarithmic function of the intensity of stimulations: $S = k \log I$, where S is the intensity of the sensation, I is the intensity of the stimulation, and k is a constant.[46] However, the intensity of the sensation caused by a stimulus of double intensity, which results from the superposition of two rays of the same colour, differs only quantitatively (by $k \log 2$)

---

45 Beyond these limits, electromagnetic waves do not give rise to coloured perceptual experiences and are therefore not visible. Rays with wavelengths just under 400 nm are called infrared; rays with wavelengths just over 700 nm are called ultraviolet.

46 The unit of measurement for S is the differential threshold (i.e., the sensation corresponding to the smallest perceived difference in intensity).

from the intensity of the sensation caused by a stimulus of single intensity. Accordingly, we need to supplement criterion (C4) with a condition guaranteeing the new quality of the emergent property.

Topological equivalence can be used as a mathematical criterion to account for the intuition of the qualitative difference between physical stimuli and sensations. The concept of topological equivalence between two spaces is formally defined by the existence of a homeomorphism between these spaces. If such a homeomorphism exists, then the spaces themselves are said to be homeomorphic, meaning that they are topologically identical. A homeomorphism is defined as a continuous bijection whose inverse is also continuous. It is not necessary to give the definitions of these mathematical terms in order to acquire an intuitive notion of topological equivalence. We can use, in fact, this intuitive criterion: two spaces are topologically equivalent if we can deform one into the other, without cutting it up or merging its different parts.

The topological non-equivalence of the one-dimensional space of physical monochromatic colours and the two-dimensional space of colour representations is manifested by a number of phenomena. The simplest of them are the complementary colours and the similarity of the colours corresponding to the stimuli at the extremes of the ordered series of physical stimuli, namely violet and red. These stimuli are separated by the greatest possible interval. In other words, violet and red are the most physically dissimilar of all the pairs of stimuli. However, their representations are no more dissimilar, for example, than the representations of yellow and green. Representations of colours located in diametrically opposed positions on the circular psychological space form complementary contrasts, giving rise to a number of well-known phenomena that have no equivalent in terms of physical stimuli: a ray composed of two rays of complementary colours of equal intensity is perceived as white (i.e., devoid of chromatic colour). Also, a small coloured surface surrounded by grey is perceived as surrounded by the opposite colour: a red disc is perceived as surrounded by green even though the background is perceived as grey if it is not juxtaposed to red.

It seems that the criterion of topological equivalence can be used to give a mathematically rigorous meaning to the idea of qualitative difference. The same criterion can be applied to physical systems. As far as the dynamic properties of a physical system (considered in classical mechanics) are concerned, a change between two dynamic states of a system is quantitative only when the trajectories of the system in its phase space — in these two states — are topologically equivalent; conversely, it is a qualitative change if the trajectories of the system are not topologically equivalent.

We can use the topological concept of novelty[47] in the definition of emergence as follows: when a purely quantitative change in microscopic parameters gives rise to a qualitative change in the trajectory of the system, this trajectory corresponds to an emergent property of the system.[48] When it is a resultant property of the system, a quantitative change in microscopic properties leads only to a topologically equivalent transformation of the system's trajectory.

However, the requirement of topological difference is too strong. The example of gases obeying the van der Waals equation shows that qualitative changes that do not correspond to topological differences can underlie emergence. Under certain conditions, these gases undergo phase transitions,[49] qualitative changes in their overall properties.

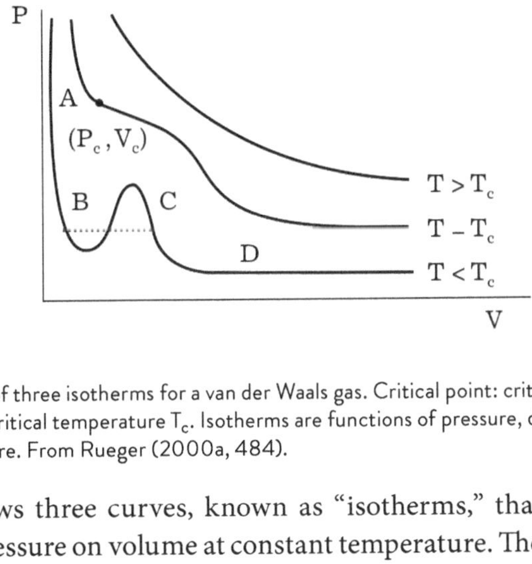

Figure 4.1 Graphs of three isotherms for a van der Waals gas. Critical point: critical pressure p_c, critical volume V_c, critical temperature T_c. Isotherms are functions of pressure, dependent on volume, at a fixed temperature. From Rueger (2000a, 484).

Figure 4.1 shows three curves, known as "isotherms," that exhibit the dependence of pressure on volume at constant temperature. The three isotherms

47 Anderson (1972) suggests using a criterion of symmetry breaking to characterize the emergence of qualitatively new properties in physical systems.

48 The trajectories of undamped and damped oscillators (Rueger 2000a) are not topologically equivalent.

49 Batterman (2002, Chapter 4) gives an accessible presentation of the reductive explanation of the phenomenon of phase transition.

shown in the diagram are qualitatively different because they do not have the same variations. Above a certain temperature, known as the "critical temperature" T_c, the volume of the gas is a strictly monotonic function of pressure; any decrease in pressure is accompanied by an increase in volume. At temperatures higher than T_c, the gas cannot be liquefied; the system remains in the gaseous state (also called gaseous "phase").

However, below the critical temperature, the relationship between pressure and volume is not monotonic. When we reduce the pressure of a liquid initially under high pressure (point *A* in Figure 4.1) at constant temperature, we reach a point (*B* in the figure) where, if we transfer heat to the liquid, the volume increases (up to point *C*) while the pressure remains constant. This process corresponds to a phase change from the liquid state to the gaseous state, which appears in the characteristic form of the isotherm for all values of $T<T_c$. The isotherms for $T<T_c$ are qualitatively different from isotherms at temperatures higher than T_c (see Rueger 2000a, 485).

We can thus reason as follows: the systemic property of being liquid or gaseous is emergent with respect to the microscopic properties of the components (the molecules) because certain small, purely quantitative, changes in the properties of the components are responsible for qualitative changes in the systemic property. A small change in the kinetic energy of the molecules (which corresponds to a change in the temperature of the gas) that causes it to pass through the critical value T_c results in a radical change in the dependence of pressure on volume. This change corresponds to the transition from a regime without phase transition to a regime with phase transition.

A consequence of my hypothesis is that a system that does not exhibit phase transitions, for example an ideal gas, has only resultant global properties: they change only quantitatively when the properties of the components undergo a small quantitative change, such as a change in kinetic energy. In contrast, a system — such as a van der Waals gas — that undergoes phase transitions has emergent global properties since these properties can change qualitatively when there is a small quantitative change in the properties of the components.

Rueger (2000a) presents systems with phase transitions as examples of "diachronic emergence." According to Rueger, a new global property, for example that of being in a gaseous state, emerges over time from another global property, that of being in a liquid state, when the pressure drops and the system is below the critical temperature. However, we can also use concepts of mathematical similarity and difference to characterize the concept of emergence of interest here. Synchronic emergence characterizes global properties with respect to the properties of the components that determine them synchronously. Unlike Rueger, I do not consider the qualitative change itself during the phase transition as a case of emergence (in the course of time); rather, I take it as a *criterion* showing that the global property that undergoes this qualitative change is emergent. One can justify this use of qualitative change as a criterion of synchronic emergence in the following way: since a quantitative change in the properties of the components determines (by a non-linear law of composition) a qualitative change in a global property, there must be a qualitative difference between the properties of the components and the global property. If the properties *C* (of the components) are only quantitatively different from the properties *G* (global), then properties C_1 and C_2, which are only quantitatively different, would determine properties G_1 and G_2, which also would be only quantitatively different. In the case of substances with a phase transition, the antecedent of the latter conditional can be true and the consequent false (in my example, if the difference between C_1 and C_2 corresponds to the difference between a state below T_c and a state above T_c); therefore, the *G* properties are not only quantitatively different from the *C* properties.

In this example, the global properties examined are dynamic properties: they are dispositional properties (powers) of the system that determine its evolution. This does not prevent them from being intrinsic properties that the system possesses at a given moment. The properties that are emergent according to this criterion, in particular the property of being disposed to a qualitatively new dynamic behaviour, are "new" compared with the properties of the components because the emergent properties are subject to different laws: they follow an evolution qualitatively different from the evolution of the isolated components by virtue of their properties and the laws that apply to them. It is the difference in the laws to which the properties are subject that proves that they are indeed different properties.[50] Accordingly, the emergent property imposes a specific evolution on the system. At each moment when the system possesses the structural conditions required for emergence (i.e., when its parts have the appropriate properties and relationships), the law of

50 In Kistler (2002a), I show that the identity of a property is determined by the set of laws in which it is involved.

composition has the nomological consequence that the system possesses the emergent property. This nomological determination is synchronic — even if the properties involved are predominantly characterized diachronically — by virtue of how their possession causes the system to evolve.

The property of a van der Waals gas of being disposed to undergo a phase transition is an emergent property: it obeys the fundamental criterion of novelty in relation to the properties of the components, a dynamic novelty accompanied by the novelty of the laws that it obeys. However, it does not obey the condition of emergence imposed by classical emergentists of being irreducible to the properties of the parts and their relationships or of being inexplicable or unpredictable. According to the concept of reduction developed above, the existence of a law of composition that determines (nomologically) that the system possesses *G* if its parts possess $P_{11} \ldots P_{nm}$ suffices for the reducibility of *G*. Once the law is known, the possession of *G* can be deduced, and consequently explained (and predicted), in the sense of the deductive-nomological model.[51]

The possibility of defining precisely what is meant by a qualitatively new property of a system, compared with the properties of its parts, allows us to add the condition of novelty to our necessary condition of emergence.

51 Rueger (2000a, 2000b) distinguishes between weak and strong emergence. A property is strongly emergent if it gives the system causal powers that the parts of the system do not possess. A property of a complex system is weakly emergent if, first, it gives the system a new dynamic behaviour and, second, it is irreducible. From the point of view of the nomological conception of the identity of properties, it seems to be contradictory to associate the novelty of the dynamic behaviour of a system with a weakly emergent property but to deny that this property gives new causal powers to the system. Only a different causal power can make the system evolve in a different way. However, Rueger's thesis, according to which even weakly emergent properties are irreducible, is only apparently incompatible with my thesis according to which emergent properties are reducible in principle. The appearance results from the use of a different concept of reduction. Rueger uses the concept of reduction common in science and analyzed by Nickles (1973). This concept characterizes a relationship between two successive theories describing phenomena at the same level, for example Newtonian mechanics and relativistic mechanics. Wimsatt calls it "successional reduction" as opposed to "explanatory reduction" (1976b, 677). In Chapter 1, I introduced the distinction between the "intralevel" successive reduction between two theories describing phenomena of the same level and the "micro-macro" explanatory reduction. An equation of relativistic mechanics, for example the equation that determines the addition of velocities, is said to "reduce" to the analogous equation of Newtonian mechanics, in the limit where the relative velocity is negligible compared with the speed of light. This sense of the term "reduce" has nothing to do with the sense that allows us to affirm the reducibility of emergent properties, which concerns the relationship between properties of different mereological levels.

(C5) A global property G of the complex object s is (strongly) emergent if and only if (it is weakly emergent and if) (1) the fact that s has G is determined by the fact that s has parts $p_1 \ldots p_n$ that have properties $P_{11} \ldots P_{nm}$, as well as by a non-linear law of composition, which is not a logical consequence of laws governing the properties $P_{11} \ldots P_{nm}$ of the parts in isolation, and if (2) the property G is qualitatively different from properties P in the sense of topological equivalence or other mathematical criteria.

The qualitative novelty of emergent properties gives rise to *stability*: the emergent property depends, of course, on the properties of the parts of the system, but it nonetheless possesses a certain autonomy that manifests itself as independence from certain microscopic changes in the parts. The fact that a system is in a solid or liquid phase is a *robust* emergent property.[52] An emergent property is robust in the sense that it has a certain (relative) independence from the properties of the components, such that a large number of small changes in the properties of the parts of the system does not induce a qualitative change in the overall property. The temperature of a gas, for example, is robust in this sense, with respect to many changes in the velocity and position of the particles that compose it: insofar as these changes — which occur continuously in the gas — do not affect the *average* kinetic energy over all particles and over time, the temperature is not affected at all and can therefore be said to be robust.[53]

52 Rueger (2000a) gives the opposite meaning to the concept of robustness, in which robust properties are non-emergent, or resultant, properties: a small change in the properties of components is reflected in a small change in the properties of the whole. For Rueger, emergent properties are therefore not robust in the sense that the non-linear nature of their dependence on component properties means that certain small changes in component properties are accompanied by (i.e., determine) radical changes in overall properties.

53 We can hypothesize that properties that are "robust" in this sense obey causal laws that are "insensitive" in the sense of Lewis (1986, 184) or "stable" in the sense of Woodward (2010). The robustness of emergent properties corresponds to the insensitivity of functional properties with respect to their different physical realizations, according to Putnam's machine functionalism. Indeed, Putnam points out that, insofar as mental states are structural functional states, they are invariant in terms of small changes in their physical realization. He draws an analogy to explanation. In his terms, "a good explanation is invariant under small perturbations of the assumptions" (1975b, 301). The good explanation refers to robust functional properties, and the assumptions refer to the physical realization. See Menzies and List (2010).

To bolster the concept of robustness as the invariance of a systemic property with respect to small variations in the properties of the components of the system, we can use the concept of *attractor* from the theory of dynamic systems. In the phase space used to represent the evolution of a dynamic system, a "fixed point" is a set of values of the variables determining the state of the system, for which the system remains stable. A fixed point is an attractor if it has a neighbourhood — called its basin — such that, if the system enters this neighbourhood, its trajectory converges toward the fixed point.

Using the notion of an attractor, we can hypothesize that each robust emergent property corresponds to an attractor, with respect to the trajectory of the system in its phase space. In this sense, the existence of an attractor can serve as a rigorous formal criterion for the existence of a robust property. Newman (1996, 2001) offers an analysis of the emergence of a property of a system in terms of its dynamics. In particular, he considers mental properties to be emergent properties of the brain, insofar as they correspond to the existence of basins of "strange" attractors in the phase diagram of the non-linear dynamic system of the brain (2001, 190). However, the condition imposed by Newman seems to be too strong: it is not necessary for the attractor to be of the "strange" type. The reason for imposing this condition is that a strange attractor causes non-periodic trajectories in phase space, which means that the system never returns to the same state twice and that two arbitrarily close points diverge exponentially in the course of the system's evolution (1996, 254–55). Newman argues that emergence requires the system to be in the basin of a strange attractor because this makes it impossible to predict the states of the system far enough into the future. But the impossibility of predicting the future states of the system seems to be neither necessary nor sufficient for emergence.

1. It is not necessary: the appearance of emergent properties, such as the ordered crystallized state during the solidification of a liquid, corresponds to a point attractor; it is perfectly predictable.

2. It is not sufficient: when the system is in the basin of a strange attractor, it does not robustly possess any systemic property. The state is sensitive to small variations in the initial conditions, whereas emergent properties are not.

The same objections can be made to a recent version of the epistemic conception of emergence, according to which a macroscopic property *P* of a dynamic system is emergent if the possession of *P* can be derived only from a microscopic description of the system and external conditions by making use of simulation (Bechtel and Richardson 1992; Bedau 1997; Holland 1998; Huneman 2008). This condition is both too weak and too strong.

1. The condition is not sufficient for emergence. A system can be complex in the sense that its state can be predicted only by simulation yet have no emergent properties. This is the case of chaotic systems that have only strange attractors, such as the double pendulum.

2. Nor is it a necessary condition for emergence. When a system has a robust emergent property, in particular a property that corresponds to the existence of an attractor, in the phase diagram of the system, that has the topology of a point or cycle, its appearance can be predicted without any simulation, simply from the presence of the state of the system in the basin of the attractor.

8. The Limits of Explaining Emergent Properties

We have seen that we can take up the vocabulary of classical emergentism only by imposing new meanings on the terms "emergent" and "resultant": none of these types of properties is determined by causal laws, and none can be discovered a priori. However, the non-causal (rather than causal) and ontological (rather than epistemic) interpretation of the concepts "emergent" and "resultant" is not the only, and perhaps not even the most important, difference between the classical concept of emergence and the concept developed here. The thesis whose refutation is generally considered to be responsible for the defeat of emergentism is that of the mysterious character of emergent properties; indeed, according to classical emergentists, the essential justification for the assertion that a property is emergent is that it is impossible to *explain* its origin.

Since Locke, the association of the perceptive experience of secondary qualities with certain stimuli has been considered as a paradigm of the mysterious character of emergent properties. Locke expresses the idea that this is a "brute nomological fact," inaccessible to any explanation, by saying that it is the impenetrable will of God, which appears to us as arbitrary, that associates a given stimulus with the smell of violets rather than with the pain caused by a steel knife cutting our flesh. He asks us to suppose

> that a Violet . . . by the impulse of such insensible particles of matter . . . causes the *Ideas* of the blue Colour, and sweet Scent of that Flower to be produced in our Minds. It being no more impossible, to conceive, that God should annex such *Ideas* to such Motions, with which they have no similitude; than that he should annex the *Idea* of Pain to the motion of a piece of Steel dividing our Flesh, with which that *Idea* hath no resemblance. (1689, Book II, Chapter VIII, para. 13)

Without using the notion of God, Alexander essentially expresses the same idea. In his words, "the existence of emergent qualities . . . is something to be noted, as some would say, under the compulsion of brute empirical fact, or, as I should prefer to say in less harsh terms, to be accepted with the 'natural piety' of the investigator. It admits no explanation" (1920, 46–47). Similarly, Broad (1925) and Lloyd Morgan (1926) insist that it is impossible to *predict* the emergent properties of a complex object on the mere basis of knowledge of its components and their properties, whereas it is possible to predict its resulting properties.

We can accept these theses of the impossibility of predicting and explaining emergent properties only if we interpret them in a new way. It is correct that we can neither predict nor explain the properties of a complex object if we know only the properties that its component parts possess in isolation from each other. However, contemporary science gives us no positive reason to think that — among chemical, biological, or psychological properties — there are emergent properties in the sense of Alexander, Broad, and Lloyd Morgan, impossible even in principle to predict and explain. The discovery of a reductive explanation, such as that of chemical bonding, of the mechanism of oxygen transport in blood or of the fixing of long-term memory refutes the idea that emergent properties cannot be explained.

Nevertheless, we can admit that the existence of emergent properties remains in a sense mysterious even if it is the subject of a reductive scientific explanation based on laws of composition. The sense of mystery stems from the fact that these laws are truths of fact rather than truths of reason. Since we *discover* the laws, in other words, because they cannot be known a priori, there is a sense in which we cannot understand why the laws have the forms that they do. When it comes to fundamental laws, we have to content ourselves with observing them without being able to hope to explain them. Ontologically, the laws of nature are necessary. However, they are epistemically a posteriori. This has the consequence that we can *conceive of* them being different from what they are. We can express this by saying that laws are "epistemically contingent."

I suggest that the persistent intuition of mystery surrounding emergent properties can be explained by the fact that their existence cannot be derived in a purely a priori way. For example, the emergence of a regular crystal from atoms that have no perceptible shape, the emergence of the shape of a flower from biological molecules, and above all the emergence of the qualitative experience of colours from the physiological mechanisms in our body continue to amaze us even when we possess a reductive explanation of these phenomena. The existence of these properties and phenomena remains partly mysterious in the same sense and for the same reason that the fundamental laws of nature remain mysterious.

Although I have just acknowledged that we can accept, in a sense, that the existence of emergent properties is partly beyond our comprehension, there is one big difference between this and classical emergentism: here emergent properties are not *absolutely* mysterious; the "natural piety," to use Samuel Alexander's (1920) expression, with which we are forced to accept their existence no longer needs to be unlimited as might have seemed to be inevitable before the advent of quantum mechanics, molecular biology, and cognitive neurophysiology. These sciences show how emergent properties can be reduced and explained. The only mystery that remains is the a posteriori nature of the laws of composition used in the reductive explanation. These laws, or at least the fundamental laws from which in turn they can be derived, always remain "brute nomological facts": even when the reductive explanation of a particular law of composition is discovered, that explanation is based ultimately on axiomatic laws that in turn cannot be explained. This observation is independent of whether the relevant law is more or less deeply integrated into a theory.[54] As a theory covering a certain domain of phenomena develops, the laws governing these phenomena change their status: what originally were "brute" empirical laws, hypothesized on the basis of induction from observation, become deducible laws within the framework of a theory. Thus, to discover an experimental law describing the behaviour of a chemical substance, at first it might be "absolutely necessary to study samples of *that particular compound*" (Broad 1925, 64). But the construction of a theory of the chemical behaviour of all compounds of a certain type can later make it possible to deduce (and in this sense explain and predict) this particular law from more general principles or laws and ultimately from the most general axioms and principles of the theory. Nevertheless, it always remains an empirical law whose explanation must end with principles and axioms that themselves are neither explicable nor predictable. For example, the law of composition that imposes the form (*) on the Hamiltonian of the molecular ion H_2^+ allows us to explain, up to a certain point, the properties of the molecule and, above all, the fact that it exists in a stable manner, thanks to the existence of a minimum of the overall energy at a certain distance R_0 between the protons. However, this explanation does not have the transparency of an a priori explanation. It necessarily starts from premises that contain an ineliminable empirical element: the particular form that the terms of the Hamiltonian take because of the interaction among the different components of the system. To distill the truth from Alexander's statement quoted above, we would have to say that emergent properties exist "under the compulsion of brute empirical law," where Alexander said "under the compulsion of brute empirical fact" (1920, 46), and instead of saying that their existence "admits no explanation" (47) it is more correct to say that it "admits no a priori explanation" or that it "admits only explanations that rest on ultimately inexplicable premises."

The concept of a "microbased" macroscopic property (henceforth MB property) developed by Kim might seem to provide a way of invalidating this result by conceiving of macroscopic properties — emergent or otherwise — as purely logical functions of underlying microscopic properties. Kim starts

54 We can put this idea parallel to the conception of explanation as resulting from the unification of the system of scientific knowledge: according to Kitcher, science explains by teaching us "*how to reduce the number of types of facts we have to accept as ultimate (or brute)*" (1989, 432).

from the concept of a "structural" property introduced by Armstrong.[55] A structural property of *s* is defined by the fact that the parts of *s* have certain properties and stand in certain relationships. Kim defines the concept of an MB property as follows: "*P* is a *micro-based property* just in case *P* is the property of being completely decomposable into nonoverlapping proper parts a_1, . . ., a_n, such that $P_1(a_1)$, $P_2(a_2)$, . . ., $P_n(a_n)$, and $R(a_1, \ldots, a_n)$" (1998, 84), where P_1 . . . P_n represent properties of the parts $a_1 \ldots a_n$ and *R* relations among these parts. This concept makes it possible to attribute to an object the property of having parts that in turn have mutual properties and relations. In this sense, it points the way to the development of the concept of a causally efficient macroproperty. However, the conditions that Kim imposes on MB properties are too weak to guarantee that the properties thus conceived are real or, in other words, causally efficient.[56] Most MB properties are not real properties in this sense (see Kistler 2005c, 149–50). Take a mereological sum whose elements do not interact with each other. The mereological whole composed of the electrons of billiard ball *A* and the nuclei of billiard ball *B* has neither the causal powers of a billiard ball nor any other causal powers. The existence of an MB property, conceived of in Kim's manner, is a *logical* consequence of the existence of the "parts" (of the mereological whole to which the MB properties are attributed), whereas the existence of a whole endowed with its own causal powers depends on the existence of appropriate interactions among the parts. Kim's definition does not impose any constraint on the relations *R* among the parts, and it does not require in particular that they are physical interactions. The subject who sees a yellow spot at a certain point in her visual field has an MB thanks to her neurons and their activation states. But even if these neurons and their activities are in their "normal" spatiotemporal relations (i.e., in conditions appropriate for producing the mental state of perceiving a yellow spot at that location), if they are prevented from interacting, then the perceptual experience disappears.

Another way of showing that the conditions that Kim imposes on MB properties do not guarantee that every MB is causally efficient is this: a given

55 "A property, *S*, is structural if and only if proper parts of particulars having *S* have some property or properties, *T* . . . not identical with *S*, and this state of affairs is, in part at least, constitutive of *S*" (Armstrong 1978, 2: 69).

56 "Real" properties are distinct from purely nominal properties that can be attributed to a system on a purely logical basis, based on the properties of its parts.

whole has a different MB for each possible decomposition. But many different decompositions do not give rise to an equivalent number of causally efficacious properties. Conversely, an object that has only one natural decomposition can have different causal powers. This is because of the interactions among the different properties of the parts. A hydrogen molecule H_2 whose most natural decomposition is into two H atoms nevertheless has several causal powers, such as its magnetic moment and a fundamental frequency of oscillation.

Unlike an MB property, an emergent property of an object is determined by a law of composition that is not purely logical and requires physical interactions among the parts. An emergent property is not identical to a structural or MB property because it is typically "multi-realizable" in two senses. First, different types of systems can possess it: the dynamic property of being disposed to undergo a phase transition can relate to the phase transition in the magnetization of an iron crystal or to the phase transition in water during freezing.[57] Second, there are many changes in the properties of the parts and their relationships that do not result in any change in the overall property of being disposed to undergo a phase transition. Many constellations of parts determine the same global property. This insensitivity of the overall property to variations in the determining properties of the parts characterizes robust systemic properties (see the previous section).

9. Avoiding Panpsychism

It seems to be difficult to deny that mental properties — as well as other properties usually taken to be non-physical, such as biological or chemical properties — are systemic. In other words, these properties appear only at a specific level: only a compound object of sufficient complexity can possess them but not its parts. Only a body more complex than an atom can be solid; only an organism can be adapted to its environment (in the sense of having fitness); only a cognitive system can represent its environment; only a human

57 Kim (1992a) shows that we can save the identity thesis in a situation of multi-realizability of this type by relativizing the global property to the different types of systems that might possess it. There would not be "the" phase transition but a different phase transition property for each type of system. However, if we consider that the identity of a property is determined by the laws in which it is involved, then we can justify the intuition that it is a single property common to different types of systems.

being can learn a systematic and creative language. To accept this idea is to adopt a doctrine that Girill (1976) calls "Democritean" in opposition to the "Empedoclean" doctrine. According to the latter, a micro-explanation of the possession of the macroproperty *P* by a system *s* must start from the fact that, among the components of *s*, some part already possesses, or some parts already possess, *P* (Klee 1984, 50). The Democritean conception, universally recognized today, admits that it is possible to explain the possession of a macroscopic property *P* of a system *s* in a way that does not presuppose that there are parts of *s* that possess *P*. For example, quantum mechanics explains the stability of molecules and the solidity of solid bodies by the stability of the bonds among atoms without attributing solidity to those atoms.

When we do not know the laws of composition that would explain a certain emergent property by a microreduction, we might be tempted to conclude that the Empedoclean doctrine is true. As far as mental properties are concerned, this is tantamount to accepting panpsychism. William James develops the argument in favour of panpsychism in this way: "*If evolution is to work smoothly, consciousness in some shape must have been present at the very origin of things*. Accordingly, we find that the more clear-sighted evolutionary philosophers are beginning to posit it there. Each atom of the nebula, they suppose, must have had an aboriginal atom of consciousness linked with it" (1890, 149).

Thomas Nagel (1979) more recently took up this argument for panpsychism, taking as premises the existence of mental properties and the impossibility of the only two conceivable ways of explaining their presence: they cannot be logically deduced from physical properties, and we have no Democritean explanation either. Therefore, we must conclude that panpsychism is true. More precisely, Nagel argues as follows.

1. Human beings are complex systems composed entirely of matter.
2. Mental properties are not logically implied by physical properties.
3. Human beings have mental properties.
4. There are no emergent properties. In other words, all properties of a complex system that are not relations

between this system and something else derive from the properties of their constituents and their mode of combination.

From these premises, Nagel concludes that panpsychism is true: "The basic physical constituents of the universe have mental properties" (181). The validity of his argument depends on the interpretation of the word *derive* in premise (4), which states that all real properties of a system can be derived from the properties of its components and how they are combined. If the word *derive* means "logically deduce," then the argument is valid.

It is well known that what is a *modus ponens* for one is a *modus tollens* for another. If the conclusion of the argument (i.e., the truth of panpsychism) is taken to be unacceptable, then one of the premises must be false. I take premises (1) and (3) to be undeniable. Thus, either (2) or (4) is false, or both are false. In Chapter 2, I critically analyzed the doctrine of "cosmic hermeneutics," according to which it is possible to deduce a priori all of the properties of the universe (including psychological properties but with the notable exclusion of qualia) from the knowledge of physical properties alone. My refutation of cosmic hermeneutics comes down to justifying premise (2) of Nagel's argument.

Now this refutation of cosmic hermeneutics also shows that (4) is false if we understand the word *derive* to mean "deduce from logical principles alone." Conversely, if the word means "deduce, possibly by means of laws of composition," then (4) simply corresponds to the physicalist requirement that I imposed at the beginning of this chapter, among other conditions, on "weak" emergence. In this case, all premises are true, but the argument is no longer valid: if mental properties are derived nomologically (4), without being derivable logically (2), then humans can be composed solely of matter (1), while having mental properties (3), without the panpsychist conclusion being true (i.e., without humans' microscopic components having mental properties).

10. Response to a Version of Kripke's Argument against the Identity Theory

According to the conception developed in this book, mental properties are global properties that emerge from the properties of the parts of the body of their possessor and from the interactions among those parts. Global properties, and mental properties in particular, are determined by laws of composition. Since these laws are necessary, it is also necessary that any individual who possesses the physical configuration underlying a mental state possesses the mental state determined by the relevant law of composition.

Kripke (1972) developed a famous argument against the thesis of the identity of a phenomenal mental property, such as pain, with an underlying physical property. Kripke began with the thesis that identity statements expressed with rigid designators are necessary and then argued that the relation between phenomenal mental properties and underlying neurophysiological properties is really, not just apparently, contingent. Now we can construct an argument analogous to Kripke's that seems to refute my conception of the relation between physical and phenomenal properties.

According to my approach, pain is an emergent property of persons and animals, determined nomologically by the interaction of certain parts of their organisms according to their properties.

> (*) Pain (*S*) = the global property *G* of the organism, determined by the law of composition *L*, from the physical property *P* (an MB macroproperty of the organism).

To simplify the analysis, let us assume that pain is not multi-realizable. In other words, let us assume that only *P* gives rise to *S* by virtue of the law *L*.

Here is an argument analogous to Kripke's. (1) If the identity (*) is true, then it is necessarily true because "pain" and "*P*" are rigid designators. (2) (*) appears to be contingent, but (3) this apparent contingency cannot be explained by a confusion between metaphysical modality and epistemic modality. According to Kripke, such a confusion explains in particular that

> (**) heat = kinetic energy

seems to be contingent. Indeed, we confuse this identity of the properties themselves (which is necessarily true) with the contingent identity of

> (***) what appears to be hot to us = kinetic energy,

where "what appears to be hot to us" is a non-rigid designator that can designate different properties in different possible worlds.

For this explanation to work, (**) and (***) must be modally different. In the case of pain, the analogue of (***) is

> (*') what appears to us as pain = the property G that P determines as a function of L.

According to Kripke, there is no difference between (*) and (*') since pain is *essentially* what appears to us as pain.

The apparent contingency of (*) therefore cannot be explained by a confusion analogous to the confusion between "heat" and "what appears to us as heat." Therefore, (*) is really contingent; (*) is not necessary; therefore, (*) is not true.

The conception of mental properties developed in this book allows us to reply to this argument as follows: both (*) and (*') are necessarily true. But (*) and (*') do not appear to us as contingent: what appears to us as contingent is the fact that it is P and the law L, or (in Kripke's original presentation) the activation of C fibres, which determine G and therefore pain in us. This appearance is justified if G is multi-realizable. G, and therefore pain, are not identical to the microbased property P that determines them. If pain is multi-realizable, then different properties P_1, P_2, et cetera can determine nomologically the same global property G. If G is multi-realizable, then it is contingent that G emerges in the human species from P_1 and not from P_2. But even if G is not multi-realizable, it is not identical to the microbased macroproperty P, which determines it through the interaction of the microscopic components of the organism. If G is not multi-realizable, in other words if G, in terms of the laws of nature, can emerge only from P, then P gives rise to G in a necessary way. The appearance of contingency is then simply the appearance of the contingency of the laws of nature (see section 4.7).

11. Emergence, Reduction, and Supervenience

Reducibility is typically understood to be the opposite of emergence. What is not emergent is often called "reducible," whereas what is irreducible is often called "emergent."[58] This equivalence is plausible only if we give the concept of reduction a very narrow meaning. According to this "simple notion of reduc-

58 This conception is implicit in Wimsatt, who distinguishes between the situation in which "the phenomena of each [level] is [sic] explained by and reduced to those of the level below" and the

tion," reducing the property of a whole consists of explaining it by "studying the parts in isolation" (Holland 1998, 14) without taking into account their interactions. I have not found it appropriate to retain this meaning, which underlies expressions such as "this conception is reductive" in the sense of "it is simplistic" or "it constitutes an oversimplification." However, as soon as we construe reduction in terms of interactions among the parts of systems, emergence becomes compatible with reduction, whereby "*an emergent property is — roughly — a system property which is dependent upon the mode of organization of the system's parts.* This is compatible with reductionism" (Wimsatt 1996, S373).[59]

Whereas reduction is often (wrongly) taken to be incompatible with emergence because the constraints that it imposes on explanation seem to be too strong, supervenience is often considered a necessary but not sufficient condition for emergence. Davidson (1970) suggested analyzing the relationship between mind and body in terms of supervenience. Supervenience is a form of correlation between mental and physical properties that appears to be sufficiently weak to be compatible with the autonomy and irreducibility of psychology. At the same time, supervenience seems to justify the physicalist thesis that the mental depends on the physical, and that the physical determines the mental, whereas the mental does not determine the physical.

However, as Kim (1990, 1993a, 1997a, 189) has shown, the concept of supervenience is independent in fact of dependence and determination.[60]

situation in which a phenomenon "might not have an explanation and thus be emergent" (1976a, 252, 253).

59 On the distinction between the widespread concept of emergence according to which emergent properties are ipso facto irreducible, and the concept of reduction adopted here according to which emergence and reducibility are compatible, see also Wimsatt (1986) and Kistler (2007).

60 Charles (1992), Horgan (1993), and McLaughlin (1995) have also emphasized the weakness of the supervenience relation, which asserts only the systematic covariation of the properties in the two sets. Charles explains that supervenience entails neither explanatory priority nor the fact that the supervenience base is the ontological basis of the supervenient properties. He points out, for example, that in a deterministic world where *S* always has effect *T*, and where *T* can only be caused by *S*, all properties that supervene on *S* also supervene on *T*. This shows that supervenience does not allow one to identify "the appropriate basis for the occurrence of a given mental property" (Charles 1992, 275). As McLaughlin points out, even strong supervenience "does not imply *explanatory* connections between supervenient and subvenient properties" and that, "*if* reduction is an explanatory relation, then the SS_m [strong supervenience, as defined with the modal operator of necessity, in McLaughlin 1995, 25] of *A*-properties on *B*-properties with metaphysical necessity fails to suffice for reduction" (1995, 48). Humphreys draws the conclusion that "supervenience does not

Supervenience does not guarantee that mental properties depend on physical properties or that physical properties determine mental properties. On the contrary, it is compatible with parallelism or occasionalism: in these doctrines, mental and physical properties are determined by an independent cause, and depend only on it, namely the will of God.

This shows two things.

1. The existence of a universal correlation between mental properties and physical properties, and even a necessary correlation as in strong supervenience, contains no indication of the origin or explanation of this correlation (see Horgan 1984; Kim 1990, 26–27, 1998, 9–15; Horgan 1993, 577 ff.).

2. The thesis that mental properties supervene on physical properties does not guarantee a materialist position. As Horgan puts it, "the mere supervenience of higher-order properties and facts on physical properties and facts cannot be enough to confer materialistic respectability" (1993, 565). What would make the explanation of the relationship complete, while showing its materialist character, would be the demonstration of a relationship stronger than supervenience (Horgan calls it "superdupervenience"), which would "constitute a kind of ontic determination which . . . confers materialistic respectability on higher-order properties and facts" (566).

The concept of emergence in terms of laws of composition, themselves grounded in laws of interaction, aims at bridging that gap. It aims to identify the nature of the determination of the properties *P* of an object *s* by the properties of the components of *s* and their relations by virtue of non-causal laws of composition. This allows us to understand why the properties *P* arise from the properties of the components, whereas supervenience is usually seen as

provide any understanding of *ontological* relationships holding between levels. For that emergence is required" (1997b, S341).

a fundamental relationship that cannot be explained.[61] Emergence implies supervenience but goes far beyond it by providing an explanation of its origin.[62] Moreover, the thesis that mental properties are emergent is incompatible with parallelist and occasionalist doctrines.

This line of reasoning allows us to accept Van Cleve's thesis that emergence is "a species of supervenience" (1990, 222).[63] But my conception reverses the order of explanation. Van Cleve proposes, in the tradition of the contemporary philosophy of mind, to start from the relation of supervenience as a genus and then to define emergence as one of its species. He proposes using the type of modality as the specific difference: whereas the definition of supervenience contains a general operator of necessity, he proposes to conceive of emergence as the species of supervenience in which the necessity is nomological and not logical. "If *P* is a property of *w*, then *P* is emergent if *P* supervenes with nomological necessity, but *not* with logical necessity, on the properties of the parts of *w*" (222).

Van Cleve's proposal shares with mine the main idea of conceiving of emergence in terms of laws of nature, but his way of specifying this idea poses two problems (O'Connor 1994, 96–97). First, if we interpret Van Cleve's distinction ontologically (and not epistemically), it presupposes the widespread but controversial doctrine that the laws of nature themselves are contingent. On the ontological level, the only way to conceive of a difference between these two types of supervenience is in terms of modal strength. Nomological necessity is supposed to be weaker than logical necessity in the sense that it only

61 In Kim (1998), supervenience provides the starting point for philosophical reflection on the relationship between mental properties and underlying physical properties.

62 We could express the relationship between these concepts by inverting Blackburn's formula according to which "supervenience is physically fixed emergence" (1993, 233). According to the conception developed in this book, emergence is a form of supervenience that obeys physical constraints: emergent properties are determined by laws that do not necessarily remain "nomological danglers." In other words, these laws have the potential to become theorems, derivable from the laws of interaction that relate to the properties of the emergent basis. In this sense, all emergent properties obey physical constraints. However, the relation of emergence is stronger than the relation of supervenience; it implies it, whereas supervenience is compatible with irreducible bridge laws and even with brute extensional correlation in the absence of any law.

63 Van Cleve (1990, 224) cites Webster's unabridged dictionary of English as evidence of the current usage in 1960, according to which supervenience was considered synonymous with emergence. The entry reads thus: "Supervene 2. *Philos.* To occur otherwise than as an additive resultant; to occur in a manner not antecedently predictable; to accrue in the manner of what is evolutionally emergent."

constrains possible worlds that share the laws of the actual world. However, the latter — the worlds that share our actual laws — are a strict subset of all possible worlds (and the associated necessity is weaker than the logical necessity) only if the laws are contingent. But there are reasons to think, on the contrary, that laws are necessary because they determine the identities of properties (see Shoemaker 1980, 1998; Kistler 2002a, 2005a). However, if laws are necessary, then there is no difference between logical and nomological necessity in terms of modal force — in other words, in terms of the set of worlds within their scope. Nevertheless, it could be argued that there is an epistemic difference between nomological necessity and logical necessity even if they are ontologically equivalent: logical necessity is accessible a priori, whereas nomological necessity is known only a posteriori. However, this epistemic difference cannot be used to develop a conception in which the difference between emergent properties and resultant properties is *ontological*; the distinction between a priori and a posteriori is an epistemic difference that does not give the two forms of arising distinguished by Van Cleve a different modal force at the metaphysical level.

Second, if we disregard this first problem and accept the premise that nomological necessity is weaker than logical necessity, we are left with the problem that Van Cleve's criterion makes all systemic properties of complex objects emergent. As we have seen, no property of a system can be deduced a priori from the properties of its components. On the contrary, they are all determined by laws that can be known only a posteriori. Therefore, the necessity with which a compound possesses them, given its components, is always nomological and not logical.

I have made the determination of emergent properties of complex objects from the properties of their parts and their interactions a necessary condition for emergence. Such a determination leads to the emergent properties arising from the properties of the parts.[64] I have argued that the discovery of the law of composition that gives rise to an emergent property goes beyond supervenience in the sense that it provides a metaphysical explanation for it. Accordingly, one might be surprised by the thesis of Humphreys that there are systems with emergent properties that violate the principle of mereological

64 My conception of cognitive properties is incompatible with Bernal Velasquez's (2012) thesis that phenomenal consciousness can have causal powers of its own only if it *does not supervene on* physical properties.

supervenience. True, Humphreys (1996, 66; 1997a, 15–16) takes the only uncontroversial example of emergent properties in this sense to be that of the properties of entangled quantum systems. However, his conception of the "fusion" of instances of properties is supposed to apply equally to cases of emergence outside quantum properties. The result of the fusion of two instances of level *i* properties, P^i_m and P^i_n, possessed respectively by the objects x^i_r and x^i_s, at time t_1, is the instance of a new property at level *i+1*, represented by $[P_{im}{*}P_{in}]$, possessed at time t_1', after the fusion, by the object $x^i_r{+}x^i_s$ resulting from the fusion of the objects x^i_r and x^i_s. One can present this formally:

$$[P^i_m(x^i_r)(t_1){*}P^i_n(x^i_s)(t_1)] = [P^i_m{*}P^i_n][(x^i_r)+(x^i_s)](t_1{'})$$ (Humphreys 1996, 60; 1997a, 9)

The time difference between t_1 and t_1' represents the fact that fusion takes time: the emergent property $[P^i_m{*}P^i_n]$ is instantiated at the instant t_1', later than the instant t_1 when the base properties P^i_m and P^i_n are instantiated by the parts of the complex object. This time lag is crucial for the conception of emergence proposed by Humphreys because it depends on the assumption that base properties *disappear* during fusion. At the instant t_1', when the fusion is completed, "the original property instances $P^i_m(x^i_r)(t_1)$, $P^i_n(x^i_s)(t_1)$ no longer exist as separate entities and they do not have all of their *i*-level causal powers available for use at the (*i+1*)st level. Some of them, so to speak, have been 'used up' in forming the fused property instance" (Humphreys 1997a, 10). The time lag between the instances of the base properties and the instance of the fused property justifies both the thesis of the novelty of the causal powers of the fused property (which in turn justifies the thesis of its emergent character) and the thesis of its non-supervenience. The instance of the fused property at t_1' has its own causal powers that are not identical to the causal powers possessed by the instances of the base properties, for the simple reason that, at time t_1', those instances of the base properties no longer exist. Given the time lag between the base property instances and the fused property, and the synchronic concept of mereological supervenience of the properties of a whole, at some instant, on the properties of its parts at the same instant, it is trivial that the fused property does not supervene on the base properties that gave rise to it.[65] Assuming that there are no other instances of the same base properties at the same place either, "trivially, there is nothing at t_1' at the i-level upon which $[P^i_m * P^i_n][(x^i_r)+(x^i_s)](t_1')$ can supervene" (Humphreys 1997a, 11).

By neglecting the distinction between the synchronic relationship of emergence and the diachronic change of properties, the conception of Humphreys trivializes emergence. In his view, the mere fact that a complex object undergoes a change that affects both its macroscopic properties and the microscopic properties of its parts is a sufficient reason to consider those macroscopic properties as emergent. According to this criterion, all macroscopic properties are emergent. Let us take a simple spatial movement. The properties of parts x^i_r and x^i_s of being at spatial positions P^i_m and P^i_n no longer have an instance at t_1'. (x^i_r and x^i_s no longer have them, and it is possible that no other object has them at t_1'.) Therefore, the property of the system being at the spatial position $[P^i_m * P^i_n]$ an instant later, at t_1', cannot supervene on any instances of P^i_m and P^i_n at t_1'.

It is true that typically (and probably always), for any emergent property possessed by an object *s*, there is an instant at which *s* begins to possess it. It is legitimate to ask for the causal process that led to this first instantiation, and it is often possible to discover it. But this question does not concern the synchronous relationship between the instance of the complex property at time t_1' and the properties possessed by the parts at that time, t_1'. To find out whether the global property *G* is emergent, we need to ask (among other things) whether the parts possess it at t_1' and whether the parts possess properties that determine it according to a law of composition. To find out whether *G* supervenes on properties *F* belonging to the parts of objects that are *G*, we need to ask whether necessarily, for any object *x* that is *G* at t_1', there exist

65 Humphreys (1997a) introduces his conception of emergence in terms of the fusion of instances of properties, in the context of the debate on the supervenience argument put forward by Kim (1998). I will return to this in Chapter 5. According to Kim, the only way in which one mental property (instance) can cause another is by causing its supervenience base. Mental causation therefore presupposes downward causation. Kim tries to refute the possibility of downward causation. The conception of Humphreys allows (i+1)-level (e.g., mental) properties to escape this argument and thus allows them to have causal powers. A property P_1 can cause another P_2 directly rather than by causing its supervenience base: in fact, P_2 has no synchronic supervenience base. But (i+1)-level properties can also have effects at level i: since P_1 does not have any synchronic supervenience base at level i either, its power to influence level i is not called into question by any such base property, whose efficacy would exclude that of P_1. See Humphreys (1997a, 14).

properties F such that the parts of x possess F at t_1' and such that any object that possesses parts with these properties F at any time t possesses G at t.

In the case of the stable state of the hydrogen ion H_2^+ that I considered earlier in this chapter, we can ask (which I explicitly have not done above) for the causal process of the fusion, over time, of two originally isolated atoms brought together until their electronic orbits overlap sufficiently for the energy levels determined by the system to have a minimum. If we consider the molecular electronic orbit that the system possesses at t_1', at the end of this process, it is trivial that it does not supervene on the electronic orbits that the isolated atoms possessed at some instant t_1 *preceding* the overlapping of the orbits: mereological supervenience — as it is imposed on emergence within the framework of physicalism — is a synchronic relation between the properties of an object and those of its parts, at the same instant. However, given that a change has taken place, after the formation of the molecule, the atomic parts of the molecule no longer possess the orbits of isolated atoms.

However, this is not enough to show that there are not other properties of these atomic parts on which the properties of the molecule supervene. If my analysis of the synchronous determination of energy states by the Hamiltonian is correct, then such properties of the molecule's parts exist: the electric charges of protons and electrons and the existence of a region of overlap of the electron orbits.

Similarly, in order to prove the novelty of a systemic property G at t_1', it is not sufficient to show that certain properties F of the parts no longer exist at t_1' and that the causal powers of G (at t_1') cannot be identical to those of F (at t_1) by consequence. Instead, it must be shown that there are no properties of the parts whose instances are synchronous with the instance of G and that have all of the causal powers of G. It follows from this analysis that Humphreys makes the mistake of concluding from the premise that there are F-properties of parts whose instances are not synchronous with the instance of G at t_1' that there are no F-properties of parts whose instances are synchronous with the instance of G at t_1'. As a result, he arrives at an extremely weak criterion of emergence according to which all systemic properties resulting from a change in the supervenience base count as emergent.

The thesis that properties at higher-level $i+1$ have no simultaneous supervenience base appears to be even more dubious in the case of mental and neurophysiological properties. At the instant t when a subject possesses mental property M, each of her neurons, synapses, and molecules possesses well-defined properties *P*. Here is a reason to contest that the *P* properties that give rise to *M* have been "used up" during the fusion that gave rise to *M*:[66] chemical and neuronal properties determine many global properties of the brain and the subject, some of which are aggregative or structural (i.e., have no causal powers of their own) and some of which are emergent (i.e., have causal powers of their own). However, if the fusion necessary for emergence made the supervenience base disappear, then the set *P* of neuronal properties could not give rise to any systemic property, not even an aggregative property. Let us suppose, *per impossibile*, that *P* gives rise to *M* as well as to an aggregative property M_1. M_1 depends for its existence on the properties *P* in its supervenience base. However, the existence of *M* requires the disappearance of *P*. It is therefore impossible for the base *P* to give rise to any aggregative property.

12. Conclusion

In light of the great scientific successes of the twentieth century, some have judged that the concept of emergence was doomed to become obsolete. According to the emergentist tradition — culminating with the work of C.D. Broad — emergence characterizes properties and nomological regularities whose existence cannot absolutely and definitively be the subject of a reductive explanation, even though these properties and regularities are the consequences of "trans-ordinal" laws that make the properties and laws of a given level depend on the properties and laws of lower levels. For emergentists, these trans-ordinal laws are absolutely inexplicable. The discovery of reductive explanations of certain chemical properties, by quantum mechanics, and of certain properties of heredity, by molecular biology, now lends credence to the conviction that there are no such inexplicable laws: the advent of the

66 This argument is due to Wong: "If all basal instances are exhausted in fusion, then structural properties and functions which depend on these will also be destroyed" (2006, 357). Wong does not solve the problem posed by Kim that I will discuss in Chapter 5: the efficacy of $(i+1)$-level properties is called into question by the underlying i-level properties. Instead of escaping overdetermination by maintaining, as Humphreys does, that the i-level properties no longer exist at the moment when the $(i+1)$ properties come into existence, Wong says two contradictory things: on the one hand, there is no overdetermination "because the basal and supervenient properties are not distinct" (357–58); on the other hand, emergent properties have new causal powers: "If basal properties *don't* possess the causal powers of emergents, then they *can't* cause the same effects; *so they can't compete as overdeterminers*" (360). It is incoherent to say both that emergent properties are not distinct from their base properties and that they have different causal powers.

reductive explanation of all the properties and laws of the macrophysical and non-physical levels of reality, starting from the microphysical level, seems to be only a matter of time.

We do not need to decide the empirical question of whether there are properties and laws that will definitively resist reductive explanation. In any case, it seems to be prudent not to base any philosophical thesis on the existence of such "emergent" properties and laws in Broad's sense. However, the concept of emergence does not lose its usefulness if we assume that they do not exist, for it can be used to account for the existence of "levels of reality." Some structured systems possess properties that are "systemic" in the sense that none of their components can possess them and are "qualitatively new." These properties are the subject of nomological regularities: that is, regularities that are not accidental but due to laws that do not exist at the level of the components. Often a particular science is devoted to the properties and laws of a given level. Chemistry, for example, studies properties and laws that are emergent in relation to those studied by physics. Emergence characterizes the relationship of determination of a higher level, such as chemistry, with respect to lower levels, such as physics. The qualitative novelty of the properties and laws of a given level is independent of the discovery of their reduction. The fact that we explain the appearance of such a property or law does not make it any less qualitatively different. The qualitative difference justifies the idea that a set of properties, linked together by laws, constitutes a distinct level of reality.

The hope of accounting for the relationship between physical and mental properties in terms of supervenience has not been realized: supervenience has turned out to be too weak to justify the belief that the underlying base properties *determine* the supervenient properties and that the latter *depend* on the former. Supervenience expresses a form of systematic and nomological correlation, but it imposes no constraint on the origin of this correlation, which makes it compatible with dualist doctrines. Emergence fills the gap left open by supervenience. It characterizes the relationship of determination between properties and laws at adjacent, but qualitatively different, levels of reality. Within the framework of physicalism, we presuppose that each level is objectively determined by lower levels and ultimately by the physical level. The discovery of the nomological form of this determination can give rise to reductive explanations (as we saw in Chapters 1 and 3), the subject of empirical discovery (as we saw in Chapter 2). Among the properties thus determined, emergence characterizes those that are qualitatively different. For example, the solidity, transparency, and redness of a ruby emerge from the properties of the atoms that make up the crystal: only a structured crystal can possess these systemic properties, qualitatively different from the properties of its components. Similarly, only a cognitive system can form representations of its environment and learn to behave appropriately in it. No single neuron — or set of neurons or even the brain cut off from the rest of the body — can represent, learn to behave, or possess properties qualitatively equivalent to such properties of cognitive systems. In this sense, representation and learning are emergent from physiological and in particular neuronal properties.

The most difficult task is to find a rigorous criterion for qualitative novelty. I have put forward the hypothesis that the relevant concept of novelty can be characterized in terms of topological equivalence and other mathematical criteria. The structure of the representations of sensory qualities, such as colours, is topologically different from that of their physical stimuli. The mathematical rigour of such mathematical criteria makes them promising. However, we are far from having specified general criteria applicable to all properties and laws that intuitively are qualitatively new. The task of elaborating such criteria is in part scientific. Philosophy can contribute to this inquiry by giving emergence a place in the conceptual landscape of the problem of the relationship between levels of reality in general and the relationship between body and mind in particular.

What remains to be done is to defend the coherence of this conception of emergence against a major objection: in order to be real, an emergent property must have causal powers of its own. However, it is doubtful that emergent properties can have such powers given that the physicalist conception of the world seems to give the monopoly of causal efficacy to physical properties underlying them. The next chapter is devoted to examining this objection.

5

The Causal Efficacy of High-Level Properties

1. Introduction

I have suggested that cognitive properties are emergent properties. According to the conception of emergence developed in Chapter 4, an emergent property is reducible in principle provided that we discover the law of composition responsible for its existence. Moreover, I have suggested a conception of reduction according to which a reducible property preserves an autonomous existence with respect to its reduction base by showing that reduction is not equivalent to the identification of the reduced property with the reducing property. But the most difficult part remains to be done. I still have the task of justifying the reality of mental properties by the causal criterion of reality: as long as I have not shown that such emergent properties can be causally efficacious, my position remains vulnerable to an epiphenomenal interpretation. To the extent that I fail to justify the ability of mental properties to make a causal difference to events in the world, their reality remains doubtful. If mental events had no causal effects of their own, then my mind would be no more than an epiphenomenon, like a shadow accompanying a real causal process taking place at the physical level.

In this chapter, I sketch a conception of causality that allows the emerging macroscopic properties of complex systems to be causally efficacious. Their efficacy complements — in a sense yet to be specified — the efficacy of the physical properties of their parts, which seem to monopolize causal power. Jaegwon Kim (1998) argues that there are only two coherent ways of doing justice to the intuition that the mind exerts an influence on its physical environment through the body's movements. Either one accepts one of the forms of dualism according to which certain causes are non-physical, for example persons,[1] or one accepts materialism, in its reductionist or eliminativist form, according to which only physical properties possess real and proper (non-derivative) causal powers and in which mental properties are understood to have, at most, derivative causal powers.[2]

According to reductionist materialism, only properties that can be reduced to physical properties are real and causally efficacious. As far as mental properties are concerned, the cognitive neurosciences of the future will show whether they can be reduced to neurophysiological properties and processes or not.[3] This makes two materialist positions conceivable: reductionist materialism (also known as "type physicalism") is the appropriate position if mental properties turn out to be reducible, and eliminativist materialism (eliminativism) is the appropriate position if they turn out to be irreducible. Now, according to Kim, a materialist (or physicalist[4]) who wishes to avoid dualism cannot maintain that mental properties are both irreducible and causally efficacious. Accordingly, he proposes that anti-reductionist materialism[5] is an unstable position that cannot be developed coherently without leading to the adoption of dualism, reductionist materialism, or eliminativist materialism.[6]

One of the aims of this book is to show that anti-reductionist materialism is not the only possible position that avoids the radical positions of dualism and eliminativism. We will see that it is not necessary to find an argument for the impossibility — in principle — of reducing mental properties to physical properties in order to avoid reductionism (understood as identification) or

---

1 This thesis, inspired by Strawson's (1959) concept of a person, has recently been defended by Lowe (2001b).

2 Among many others, Armstrong (1968) and Kim (1998) are materialists in this sense.

3 This statement is supposed to apply to the vast majority of cognitive properties that have not yet been the subject of a reductionist explanation. We saw in Chapter 1 some examples of cognitive properties that have already been reduced.

4 Kim (1998, 2, 2005) speaks of physicalism rather than materialism. I take these terms to be equivalent, both expressing the doctrine that (1) every object is composed exclusively of parts that have only physical properties, and (2) every intrinsic property of a complex object can be reduced in principle to the physical properties of its parts. It seems to me that there are good reasons to subscribe to the first thesis, the truth of the second thesis being an open empirical question.

5 This position became very influential following the writings of Putnam (1967) and Fodor (1974).

6 Of non-reductive physicalism, Kim says that "this intermediate halfway house between the two poles of substance dualism and reductionist physicalism is a promissory note that cannot be redeemed" (2005, 158).

eliminativism. It is up to science, not philosophy, to discover which mental properties are locally or globally reducible to neurophysiological properties, possibly through the construction of new psychological or neurophysiological concepts. But questions concerning their reducibility are independent from questions concerning their causal efficacy. The causal efficacy of a property is not threatened by a reduction (as dualists and some reductionists claim), and a reduction is not necessary to ground it (as some physicalists claim). Of course, the reducibility of a property plays an important role in our *understanding* of its causal efficacy. The reduction of a cognitive capacity, for example through the discovery of a mechanism underlying its exercise,[7] is the best way of explaining why and how this capacity exerts a causal influence. However, ontologically, its efficacy depends not on its reducibility but on the laws of nature that link it to other properties at the same level: that is, properties of the entire cognitive system and not of its parts.

In what follows, first I will sketch a conceptual framework within which the question of the causal efficacy of mental properties can be posed without prejudging it in the direction of reductionism, eliminativism, or dualism. Next I will show that this framework allows us to defend the possibility of the causal efficacy of the mind against two important objections: first, Kim's objection that mental properties can have only derivative efficacy; second, Lowe's objection that such efficacy can be acknowledged only within a dualistic framework.

I will ignore the more specific problem of the intentionality of mental states and properties. At least some mental states possess a *content* constitutive of their identities. According to the externalist conception of content, the identity of such an intentional mental state is determined by things too remote in space and time from the person in the mental state in question to be able to contribute causally to the efficacy of the mental state. Therefore, we are faced with the problem of understanding how mental states can be efficacious by virtue of their content. However, this problem can be approached independently[8] of the problem analyzed here: is it conceivable and plausible

7 On the notion of mechanism, see Glennan (1996, 2010); Machamer, Darden, and Craver (2000); Craver (2007); Craver and Darden (2013).

8 One promising approach is the strategy adopted by Dretske (1988) to distinguish between bodily movement and behaviour and their respective explanations. Dretske construes behaviour as the process that causally brings about bodily movement. According to Dretske, content contributes causally to the evolution (in the biological sense of evolution by natural selection) of behaviour

that global properties of complex objects are causally efficacious so that eliminativism, reduction understood as identification, and dualism can all be avoided? My strategy for answering this question consists of putting the issue of the efficacy of the mind into a more general perspective. Part of the problem of mental causation is just as much a problem for the theory of macroscopic causation in general (Baker 1993, 79; 1998, 261). We can therefore hope to make progress in our understanding of the particular problem by justifying the possibility of macroscopic causation in general.

2. Causality, Causal Responsibility, and Causal Explanation

Let us take a statement expressing mental causation. The thought that street noise disturbs my concentration causes me to decide to close the window. This is a case of mental causation in the strict sense: that is, a situation in which both cause (the thought) and effect (the decision) are mental events. But common sense also naturally conceives of psychophysical causation, in which a mental event causes a physical event, for example if my decision (a mental event) causes my act of closing the window (a behaviour and therefore a physical event). Of course, mental events can also be effects of physical causes, as happens in perception: the (physical) noise in the street causes me to think that this noise disturbs me. Each of these mental events also has physical properties: thoughts and decisions always occur in humans, who are material beings.[9] Moreover, the mental properties of these events are determined

---

understood in this sense. Therefore, the content of a cognitive state can be the part of the "structuring cause" of some type of behaviour, whereas it is not involved in triggering bodily movement.

The development of the concept of narrow content is another promising line of research (Lewis 1994; Braddon-Mitchell and Jackson 1996; Chalmers 1996). The general justification for setting aside the causal role of content is that this is a more general problem, in the terms of Crane and Mellor, of finding a "local *causal surrogate*" (1990, 194) *S* for a property *P*, where *P* is a relational or otherwise extrinsic property and *S* is the property *directly* (and locally) responsible for the effects of *P*. Postulating an intrinsic mental property directly responsible for the effects of a subject's thoughts (themselves extrinsic insofar as their content is partly determined by their relations to states of affairs outside the thinker) can be seen as analogous to postulating a local electric field directly responsible causally for the acceleration of an electric charge q_1. The local intensity of the field is the local causal surrogate of the relational (extrinsic) property of q_1 of being located at a certain distance from another charge q_2.

9 My adoption of a Davidsonian terminology, according to which an "event" is a particular entity that possesses many properties, is not intended to prejudge the question that occupies us in

by underlying physical properties: if we intervene on the relevant regions of the brain, then the thought or decision can be altered or disappear. There is no doubt that the causal relationship between these events follows the laws of physics, which apply to them according to their physical properties. We can therefore wonder about the causal contributions of their mental properties: does the fact that it is a thought with a certain content (the noise in the street is disturbing my concentration) contribute causally to my decision to close the window? Does the fact that it is a decision with a certain intentional object (to close the window) contribute causally to my act of closing it? Two intuitively plausible principles appear to challenge the idea that mental properties make a causal difference. First, according to the principle of "causal closure," every physical event, at every instant preceding it, has a complete and purely physical cause. In particular, my act of closing the window, as a physical event, has a complete physical cause at the moment that I decide to close the window. Second, an event's mental properties differ from its physical properties. It seems, then, that we are led to the conclusion that the mental properties of the event, in this case the property of being a decision, are causally inert: if my act of closing the window is the result of an uninterrupted and purely physical causal chain, then there seems to be no room left for contributions of mental properties. This line of reasoning, developed with great clarity by Kim (1998), assumes that causal overdetermination is exceptional. There can be rare events caused by two independent causal processes that converge. This can be the case, for example, when someone sentenced to death is shot by two marksmen in such a way that their bullets hit her heart at exactly the same time and in the same place. In this case, her death is overdetermined: each bullet is sufficient for her death, such that the removal of one would not alter the result. The argument for the causal inertia of mental properties depends on the premise that such overdetermination is exceptional. In other words, it is not plausible to assume that, systematically, each effect of every mental cause is overdetermined, in the sense that it has both a physical cause and a mental cause. Kim expresses the thesis that overdetermination is exceptional

---

this book, that of the relationship between the mental properties and the physical properties of these events. It would be just as possible to pose the question in the alternative terminology proposed by Kim (1973), according to which an event is the exemplification of a property by an object at a given moment. In this terminology, we would have to say that each event in which I think that street noise disturbs my concentration is accompanied by (or supervenes on) an event in which my brain has a certain neurophysiological property.

in his "principle of causal-explanatory exclusion." There can be only one complete causal explanation of any given phenomenon. This line of reasoning represents a formidable challenge to the conviction of common sense that the mental nature of our thoughts and decisions has a causal impact on the world.

In this chapter, we will see how it is possible to defend this intuition against the thesis that mental properties are epiphenomenal: in other words, that they make no causal difference of their own to the course of events. I will reply to the challenge that all causes are physical and that, more specifically in the case of the physical consequences of our actions, all of their causes are neurophysiological.

Before going into the details of this debate, it is necessary to respond to an important objection to my way of posing the question of whether macroscopic properties have causal efficacy beyond that of the underlying physical microscopic properties. According to some authors, the question of whether certain properties of a cause are efficacious with respect to the properties of the effect is ill posed. In particular, Donald Davidson (1980, 1993) argues that properties (and the predicates that refer to them) belong to the conceptual register of *explanation* rather than that of causality. According to Davidson (1995), it is true a priori that, for any pair of causally related events, there is a law of which this causal relationship is an instance; therefore, it is always possible in principle to explain why one caused the other. However, Davidson rejects the very question that raises the problem of mental causation. How can a mental event cause something like bodily motion by virtue of its mental properties given that the same event also seems to cause the same motion by virtue of its physical properties?[10] Having adopted an ontological framework that gives no place to properties in the analysis of why one event causes another, Davidson is challenged by critics who argue that he "holds doctrines which commit him to denying that mental events cause physical events in virtue of falling under mental types. On his view, they claim, the mental *qua* mental is causally inert" (McLaughlin 1993, 28). Davidson's metaphysics of causation, which admits only particular events and their linguistic descriptions, does not make sense of the question of what makes a given event cause another event *by virtue of specific properties.* From his point of view, this question betrays confusion between a demand for information about a causal

10 Antony (1991) shows that Davidson must reject it given his conception of the attribution of mental properties, according to which it obeys normative constraints of rationality.

relationship and a demand for a causal *explanation*. The first is an extensional relationship between particular events. The second is a relationship between statements: one statement explains another, the *explanandum*, if it is possible to construct a deductive argument whose conclusion is the *explanandum*, and whose premises, which together constitute the *explanans*, contain the first statement together with a certain number of nomological statements.

However, the question of whether certain properties are causally efficacious is important. A satisfactory theory of causality must provide a conceptual framework for identifying what it is about the cause that is responsible for its effect having certain properties. If a red billiard ball is the cause of the fact that a white billiard ball starts moving in a given direction with a given speed, then it causes this precise movement by virtue of the energy and momentum that it carries before the impact but not by virtue of its colour. This would be a fact even if there were no science, no language to express it, or any statement of law or explanation that referred to it. Some properties of the causing event (e.g., its momentum) objectively modify the relevant properties of the effect, whereas others (e.g., its colour) do not. Therefore, a complete theory of causation must acknowledge the objective role that the former play in determining causal interactions.

We could simply ignore Davidson's nominalist scruples, which make Davidson prefer the language of predicates to that of properties and translate his position into realist terms. Instead of simply saying that events satisfy predicates, we would say that they have properties. Davidson (1993) himself uses this realist language; however, he would not accept the next step in the following reasoning. Instead of simply admitting that referring to F_1 (the fact that the red ball hit the white ball with momentum M) causally explains F_2 (the fact that the white ball, after being hit by the red ball, has momentum M), we must acknowledge that, if this explanatory relationship is correct, then its truth has an objective basis, in other words a "truth maker." We can express this by saying that F_1 is *causally responsible for*[11] F_2. In this way, we can account for the fact that the distinction between F_1 (the fact that the red ball hit the white ball with momentum M) and F_3 (the fact that the red ball was red when it hit the white ball) reflects an objective difference in the causal influence that the constitutive properties of these facts have on the motion of

11 I have developed the notion of causal responsibility in Kistler (1999a, 1999b, 2001, 2002b, 2006a, 2006d, 2014).

the white ball. This difference does not depend on an explanatory difference; F_1 but not F_3 would be causally responsible for the effect even if there was no one to seek or offer explanations.[12]

However, this argument is still not sufficient to overcome the specific obstacle that prevents, according to Davidson, mental properties from being causally efficacious. My reasoning has no force insofar as mental properties are concerned because they cannot even figure in causal *explanations*. Therefore, it cannot be said, even within a realist reformulation of Davidson's position, that these properties can participate in facts causally responsible for anything. The reason is that there are no strict laws involving mental properties, whereas causation presupposes the existence of strict laws. The answer that seems to be plausible to me is that it is not necessary for a property to fall under a strict law in order to be mentioned in a causal explanation and thus to be causally efficacious. If it is true that most laws are not strict,[13] then Davidson's version of the nomological theory of causation has the consequence that most physical properties are not efficacious either. This is a reductio of the condition that the relevant laws must be strict. Some authors maintain that the existence of a non-strict law (or *ceteris paribus* law) that applies to a given situation is sufficient for the existence of a causal relationship between the events to which the law applies.[14] However, if the condition of the existence of a strict law is too strong, then the mere existence of a non-strict law is too weak to account for causal efficacy.[15]

---

12 Putnam (1992, 47 ff.) and Hardcastle (1998) suggest reducing the ontological problem of determining which properties are causally efficacious to the epistemological problem of knowing which explanations are pragmatically preferable. This amounts to putting the cart before the horse: some explanations are objectively *more correct* than others given the *explanandum*, independent of our interests. The quantity of movement, not the colour, is efficacious in relation to the quantity of movement of the ball set in motion. The ontological relationship of causal responsibility is what makes an explanation objectively correct.

13 This thesis has been defended by, among others, Joseph (1980); Cartwright (1983); Hempel (1988); Fodor (1989); Pietroski and Rey (1995); Kistler (1999b, 2006d). Laws without exceptions are called "strict." The laws of "special" sciences (i.e., those other than fundamental physics, e.g., psychology or economics) are often considered to have exceptions, so they are "not strict." The question of which physical laws are strict is controversial. See Kistler (1999b or 2006d, Chapter 3; 2006b).

14 See Fodor (1989); McLaughlin (1989, 1993); Pietroski (1994); Antony and Levine (1997); Glennan (2010).

15 Robb (1997, 181) makes a brief remark along the lines of the argument that I will offer here against this thesis.

The following situation shows why the existence of a law expressing a dependence between certain properties of two events is not sufficient for these events to be linked as cause and effect. Let us assume that *A* is a radio station that broadcasts a certain program so that the waves carrying the signal travel in all directions at the speed of light. Let us then take two locations, *B* and *C*, situated at equal distances but in opposite directions from *A*. Let us call (*B*, *t*) and (*C*, *t*) the events of the arrival of the signal at *B* and *C*, at time *t*. There is a locally valid law that links (*B*, *t*) and (*C*, *t*); in fact, the propagation of electromagnetic waves in a vacuum follows from Maxwell's equations.[16] Therefore, according to the realistic theories of mental causation proposed by Fodor (1989), McLaughlin (1989, 1993), and Pietroski (1994), (*B*, *t*) and (*C*, *t*) should be causally related as cause and effect, which they are not. Two events (*x*, *t*) and (*x'*, *t'*) are "spatially separated," in the sense of the theory of special relativity, if they cannot be linked by a light signal. (*B*, *t*) and (*C*, *t*) are spatially separated in this sense and cannot therefore be linked by any causal process. Rather, (*B*, *t*) and (*C*, *t*) are effects of a common cause (*A*, *t'*) taking place at *A* a little earlier than *t*.

The existence of a counterfactual dependence, a condition proposed by LePore and Loewer (1987, 1989), is too weak to guarantee the existence of a causal relationship, for the same reason. The events of the arrival of the signal at *B* and *C* depend counterfactually on each other. In the imagined situation (i.e., supposing that there are no other sources of signals or screens), if no signal arrived at *B* at *t*, then no signal arrived at *C* at *t*; if none arrived at *C*, then none arrived at *B*. That counterfactual dependence is not sufficient to

16 This law is *local* because its existence depends on the transmission of a signal from *A* and the absence of electromagnetic screens that would prevent *B* or *C* from receiving the signal. But many paradigmatic laws, such as the law of free fall near the Earth's surface, have their validity locally limited in a similar way. The nature and logical structure of the laws of special sciences are the subject of a debate in contemporary philosophy of science. See, among others, Cartwright (1983, 1989); Pietroski and Rey (1995); Earman and Roberts (1999); Schurz (2002); Kistler (2006b). This is not the place to go into the details of this debate. But the conclusion that emerges is that the laws of the special sciences are "system laws" (Schurz 2002) that apply to specific systems over limited time intervals. When we determine the systems to which these laws apply — they are, to use an expression introduced by Cartwright (1989), "nomological machines" — we generally mention both properties that belong to the level of the law and properties that belong to lower levels. For a physiological law concerning the exchange of gases between the lungs and the blood, which applies to a particular biological species, there is a set of physical, chemical, and biological conditions that characterizes the type of system to which the law applies.

establish that one of these effects is the cause of the other.[17] Indeed, here it is clear that this is not the case.[18]

I propose to draw the following lesson from the case of the radio transmitter. Causality, nomological, and counterfactual dependence are different concepts without being independent. In particular, neither the causal relationship nor the relationship of causal responsibility can be reduced directly to nomological or counterfactual dependence or to increased probability.[19] Neither nomological dependence, nor counterfactual dependence, nor an increase in probability between two types of events (or between properties of events) makes it possible to deduce conclusively that a particular event belonging to one type is the cause of an event belonging to the other type. All of these relationships exist between types, or sets, or properties of events. But we cannot draw any conclusion about the existence of a causal relationship between two *particular* events from the existence of one of these relationships at the level of types.[20] Therefore, what makes *c* cause *e* must be a fact about the particular events *c* and *e* and the relationship between them as particulars.

---

17 The situation that we have just considered is compatible with Carroll's assertion according to which, if two events are counterfactually dependent on each other, they "*belong to a single causal network*" (1994, 121). However, this condition is far too weak to allow us to discover what is causally responsible for what. In fact, it seems to be reasonable to assume that all events that have ever taken place or will ever take place at any time in the entire universe belong to a single causal network.

18 The concept that Horgan called "quausation" between a fact about the cause and a fact about the effect is also defined in terms of counterfactual dependence. However, Horgan's "quausation" does not, at first sight, appear to be an objective relationship at all insofar as it contains the relationship of being "explanatorily relevant" (1989, 50). What is relevant to explain a given *explanandum* depends on the pragmatic circumstances, in particular the prior knowledge and interests of a person who has asked for the explanation and listens to the response. True, Horgan claims that "*bona fide* quausal relevance is not merely epistemic, but metaphysical" (1989, 53–54). In this case, it seems to be mistaken to try to define it in terms of the pragmatic notion of relevance.

19 Situations such as the one that I have described, therefore, refute the traditional nomological analysis of causality, developed by Hempel and Oppenheim, Popper, Carnap, and others, but also Lewis's (1986) analysis in terms of counterfactual dependence and the analysis according to which causality is equivalent to an increase in the probability of the occurrence of an event of a given type given the occurrence of an event of another type (Eells 1991). The theory of causation in terms of interventions (Woodward 2003) also faces difficulties (Kistler 2013). For brief presentations of all these approaches, as well as critical remarks, see Kistler (2002c, 2004a, 2011, 2025); Schaffer (2014).

20 Eells expresses this clearly with regard to the analysis of causation in terms of increased probability: "Given the conceptual independence of token-level causal facts from type-level causal facts, it should not be surprising that what is true at the type level is . . . conceptually independent

This fact is added to relationships existing between their properties, such as nomological or counterfactual dependence or probability increase. This is not the place to defend the thesis that the transmission of an amount of a conserved quantity between two events is a necessary and sufficient condition for the existence of a causal relationship.[21]

Let us now return to the question of what makes certain properties of an event efficacious in a given causal relationship while other properties play no such role. The concept of causal responsibility (CR) can serve as a framework for asking this question. It can be analyzed as follows.

> (CR) The fact that *c* is *F* is causally responsible for the fact that *e* is *G*, if and only if *c* is a cause of *e* (at the level of particular events, by virtue of transmission of conserved quantities), *c* exemplifies *F*, *e* exemplifies *G*, and there is a law (which generally is not strict) according to which instances of *F* tend to produce instances of *G*.

In (CR),[22] the expression "*c* is the cause of *e*" designates a causal relationship between particular events based on the transmission of a certain amount of a conserved quantity. An event is a particular whose identity conditions are given by the limits of the spatiotemporal zone that it occupies. The statement "the fact that *c* is *F* is causally responsible for the fact that *e* is *G*" implies both that *c* is the cause of *e* and that the property *F* of *c* is causally efficacious in the production of an event that has property *G*.[23]

To guarantee that causal responsibility, which depends on properties, is as local as the causal relationship between events, we need to conceive of the properties *F* and *G* that constitute the facts that *c* is *F* and *e* is *G* as instances

---

of what is true at the token level, and that token-level causation cannot be straightforwardly understand [sic] in terms of type-level causal relations" (1991, 16).

21 See Kistler (1998, 1999b, 2006d). The proposal to reduce the causal relationship to one of transmission between events contains a Davidsonian element. In this analysis, the terms of the causal relationship are conceived of as particular Davidsonian events. Events thus construed have many properties, only some of which are mentioned in the expressions used to refer to them.

22 Causal responsibility has similarities to what some authors have called "qua-causation" or "quausation." See Horgan (1989); McLaughlin (1993); Kim (1993c); Marras (1998).

23 Since many events are "temporal parts" (or "temporal slices") of objects, what I call "facts" can be considered as a category that contains, as a special case, what Kim (1973) calls "events."

of properties or "tropes" and not as universals.[24] Only a spatiotemporal entity, such as a trope, can be locally efficacious, whereas universals are not localized in space and time.[25] According to an Aristotelian conception of universals, we could say that a universal is localized wherever its instances are localized. But then what is efficacious in a given situation is an instance, not the universal itself, the totality of all instances. This is just another way of saying that the efficient entity is a trope. However, this idea itself is not enough to solve the problem since the crucial question now becomes that of identifying causally efficacious tropes. Under which conditions is one trope identical to another or distinct from another? Our central question thus becomes this: if two tropes of the same event are not identical, then which one was efficacious in relation to a given effect? Robb (1997) proposes the following solution: if I decide to close the window, at the moment of the decision I have a mental trope identical to a physical trope. This decision trope is mental insofar as it belongs to a mental type of decision to close the window, but it is also physical insofar as it belongs to a certain physical and more specifically neurophysiological type. However, it is causally efficacious not by virtue of the fact that it belongs to one type or the other: it owes its efficacy neither to the fact that it belongs to the mental type nor to the fact that it belongs to the physical type.

The identity conditions for Robb's tropes are less coarse than the identity conditions for Davidsonian events.[26] But they are similar with respect to their "coarseness" in a sense that I will specify in a moment. For this reason, Robb's solution is ultimately very similar to Davidson's. Robb's tropes are "coarse" in the sense that they can be both mental and physical, just as Davidsonian events. Mental tropes are causally efficacious because they are identical to physical tropes, assumed to be causally efficacious. In this statement, we need only substitute the word *event* for all occurrences of the word *trope* to re-

24 Macdonald and Macdonald (1986, 37–40), Heil (1992, 136–39), Ehring (1996), and Robb (1997) have defended this thesis specifically for the case of mental causation. Keith Campbell advocates it for causation in general: "The terms of every real causal sequence are one and all of them particulars. When you drop it, it is the weight of this particular brick, not bricks or weights in general, which breaks the bone in your particular left big toe" (1990, 113). See Kistler (1999b, 2006d).

25 Of course, universals are said to be "wholly present" in each of their instances, and instances are indeed located in space and time. But an entity "entirely" present in innumerable places and at innumerable times is not spatiotemporal in the sense required for causal efficacy: in that sense, it must act at a particular place and time, to the exclusion of other places and times.

26 See Davidson (1980), especially the essay "Mental Events."

trieve the Davidsonian solution to the problem of mental causation. Tropes sufficiently coarse to be both mental and physical prevent us — just as much as Davidsonian events — from asking the question of what it is about these events/tropes that is causally efficacious in bringing about a given effect.[27] Robb's tropes, just like Davidson's events, lack the internal structure that would make it possible to distinguish between different *aspects* that could bear different causal responsibility. Davidson's anomalous monism is confronted with the objection that it renders the mental epiphenomenal (see Sosa 1984; McLaughlin 1993). It prevents us from giving an affirmative answer to the question of whether a given event, for example the decision to open the window (which caused me to open the window), caused the latter act by virtue of its mental properties (the fact that it was a decision) or whether its efficacy is due exclusively instead to its neurophysiological or physical properties. Robb's theory, which uses the concept of trope, similarly precludes the question of whether the mental and physical trope caused my act of opening the window by virtue of its mental aspect or by virtue of its neurophysiological aspect.

We can hope to answer questions of this kind only on the basis of a criterion of the identity of properties (or tropes).[28] The following nomological criterion links the identity of a property (trope) to its nomological relations with other properties (tropes of other types).

> (Nomological criterion of property identity) Property *P* is identical to property *Q* if and only if, for all properties *R*, *P* is in a nomic relation *N* with respect to *R* if and only if *Q* is in the same nomic relation *N* with respect to *R*.[29]

27 Noordhof (1998) makes a similar criticism of Robb's proposal, as do Yablo (1992, 259n32) and Lowe (1993, 631; 1996, 74) with respect to the proposals by Macdonald and Macdonald and Heil, to analyze mental causation in terms of tropes. Robb anticipates this criticism by pointing out that allowing *aspects* of tropes to be causally efficacious would lead to a vicious regress. However, the conclusion to be drawn is that it is necessary to find a different way of justifying the causal role of the mental as opposed to the physical. If it is impossible to use aspects of tropes for this purpose, then we must distinguish between the tropes themselves.

28 The criterion is supposed to apply to properties as such independently of their conception as universals or tropes.

29 Achinstein (1974) proposes a similar criterion for the identity of properties, in which causal equivalence plays the role assigned here to nomic equivalence. Achinstein's criterion would not be appropriate for my project. Since it is concerned with causality between events, Achinstein's

Formally: $\forall P \forall Q \{P = Q \leftrightarrow \forall R \forall N [N(P,R) \leftrightarrow N(Q,R)]\}$

P and *Q* are identical properties if and only if they share all of their nomic dependency relationships with respect to other properties; moreover, since causal responsibility is determined by nomic dependence, identical properties enter into the same causal responsibility relationships.

Let us consider once again the example of the red billiard ball that causes the movement of a white billiard ball, with a given speed and in a given direction. There are laws linking energy and momentum, and therefore the speeds of these two balls, but there are no laws linking their colours and their speeds. Consequently, the colour trope of the red billiard ball is different from its speed trope, and what is causally responsible for the speed of the white ball after the impact is the speed and not the colour of the red ball.[30]

According to this criterion, mental and physical properties are different: they are involved in different laws. In this framework, the crucial question becomes whether there are psychological laws between different mental properties. If such laws exist, then mental properties are causally efficacious. This is an empirical question.[31]

I propose to construe a mental property *M* as a first-order macroscopic property of an individual *s*. *M* is determined by a complex physical property *P*. *P* is constituted by the logical conjunction of the local properties of the bodily parts of *s* as well as the relationships among these parts. *P* determines *M* in a nomological but non-causal way. The properties of the parts of *s* and the laws governing the interactions among these properties together give rise to a law of composition. Each individual *s* that has parts $s_1 \ldots s_n$ with properties

criterion cannot distinguish between the mental and physical properties of a given event. In any case, the relevant nomic relationships, which might allow us to distinguish between neurophysiological and mental properties of a single event taking place in the brain-mind, cannot be interpreted as causal because causation requires its terms to have distinct spatiotemporal locations.

30 We saw above that the fact that a law applies to a property (trope) in a given situation, and the fact that this property is counterfactually dependent on another, are not sufficient conditions for the existence of a causal relationship between the facts involving these properties. Nevertheless, they are necessary conditions. The laws to which a given trope belongs determine the identity of that trope, but the laws also determine which of several tropes instantiated in the same event is causally responsible for a given effect.

31 We have already seen some examples of such laws. In this chapter, I take up the example of Rescorla and Wagner's law of classical conditioning. Crane and Mellor (1990), Antony and Levine (1997), and Rey (1997) have also offered arguments for the existence of psychological laws.

$P_{11} \ldots P_{nm}$ possesses, by virtue of the laws of nature, global properties $G_1 \ldots G_r$, some of which are mental. Similarly, the microphysical state of a gas, conceived as the conjunction of the properties of position and momentum of the molecules that compose it, naturally determines, by virtue of a law of composition, the overall temperature of the gas.

Non-causal nomological determination is distinct from two other recent conceptions of the relationship between a person's mental properties and the underlying physical properties of her body.

First, according to the functionalist view, mental properties are functional properties. Attributing to someone the property of feeling pain is equivalent to attributing to her a property that satisfies a functional condition, in other words a second-order property expressed by means of an existential quantification. In the simplest case, this is a condition expressed in terms of sensory causes and behavioural effects. To say that *x* feels pain is equivalent to saying that *x* has one or another of a set of neurophysiological properties that are caused by bodily harm and that in turn cause (among other things) behaviour that leads to an escape from the source of the harm. The cause of this escape behaviour, which at the same time is the effect of the external stimulus, is a first-order neurophysiological property said to "realize" the mental property in that individual. However, according to this view, the mental property itself is causally inefficacious; in other words, it is epiphenomenal. Provided that we accept the causal criterion of reality,[32] this has the consequence of denying any reality to the mental property beyond the physical property that realizes it. What is properly mental is merely a second-order *predicate* constructed by an existential quantification over first-order physical properties; only the latter are real and causally efficacious.[33]

Second, according to another proposal (Yablo 1992), the relationship between mental properties and underlying physical properties can be likened to the relationship between a determinable property and one of the properties that determines it. The mental property would be to the underlying physical property what red is to scarlet or what temperature is to 0° Celsius. Indeed,

32 According to the causal criterion of reality, the fact that an entity is capable of making a difference to causal interactions is a necessary and sufficient condition for judging it to be real. See Kistler (2002a).

33 Kim (1998) explicitly draws this consequence from the functionalist conception and accepts it.

it seems to be plausible that the mental property of feeling pain is an abstract and determinable property, whereas each biological species capable of feeling pain, and perhaps even each individual capable of feeling pain, has its own specific painful sensations.

However, the relationship between determinable and determinate is inappropriate as a model for the relationship between mental properties and underlying physical properties. A first reason is that the different physical properties on which the existence of a given cognitive property depends in different species and individuals are not ordered in a series, as is the case with the set of determinates of a determinable: the set of temperatures, for example, is ordered in one dimension. It is not clear how this model could be extended beyond the domain of quantitative properties, where more or less fine discriminations give a clear meaning to the distinction between properties more or less abstract and therefore more or less determinable.[34]

The second and most important reason for considering that the model of the distinction between determinable and determinate is not appropriate to the analysis of the relationship between mental properties and physical properties is that, to be in a relationship of determinable to determinate, two properties must be exemplified by the same object and be of the same logical type. This is the case if both the determinable and the determinate are mental properties, for example the general property of experiencing pain and the human property of experiencing pain. The whole individual possesses both properties. Conversely, the physical property underlying the mental property of experiencing pain, although it too can be technically attributed to the whole individual, is in reality the conjunction of properties of *parts* of the individual (of nerve cells, neural circuits, or brain regions) as well as of relationships among these parts. In Kim's terminology, this underlying physical property is a "micro-based property" (see Chapter 4.7) that "belongs to a whole in virtue of facts about its parts" (1988b, 142; 1993b, 124) and by virtue of logic alone: "*P* is a *micro-based property* just in case *P* is the property of having proper parts, a_1, a_2, . . ., a_n, such that $P_1(a_1)$, $P_2(a_2)$, . . ., $P_n(a_n)$, and $R(a_1, \ldots, a_n)$" (1997b,

34 Funkhouser (2006, 565) and Menzies (2008, 203) criticize Yablo's thesis within the framework of the analysis of the determinable/determined relationship proposed by Funkhouser. A given mental property can be "superdetermined" (i.e., correspond to a point in the space corresponding to its "determination dimensions") yet be realizable by different physical properties. In this case, the physical realizing property cannot be a determinate of which the mental property would be a determinable, since the mental property itself is already maximally determinate.

292). The simplest case of a micro-based property is the property of "being made up of two parts *x* and *y* such that *x* is *F* and *y* is *G* and *x* is related by *R* to *y*" (Kim 1988b, 142; 1993b, 124). Now the mental property of feeling pain is determined by the properties of the parts of the organism and their interactions, *by virtue of a law of nature*, and not just by virtue of logic. For this reason, it is not a micro-based property in Kim's sense. Since the mental property is not micro-based, whereas the underlying physical property is, these two properties belong to fundamentally different kinds, making them incapable of being in the relationship of determinate to determinable.

This point can also be expressed as follows: the relationship between a determinable and a determinate is an *internal* relationship. According to a plausible conception,[35] a determinate is a complex property: it is designated by a complex predicate that has the form of a conjunction. When one of the terms of the conjunction is deleted, the resulting predicate designates a determinable property relative to the determinate designated by the original predicate. There is then an internal relationship of subordination between these properties: any object that has the determinate also has the determinable because the elimination of certain terms of the conjunction corresponds to a valid inference. Furthermore, if we know that an individual *s* possesses the determinate property, and if we know the conjunctive structure of this property, then we can infer a priori, for purely logical reasons, that *s* also possesses the determinable. This model is unsuitable for analyzing the relationship between mental properties and underlying physical properties insofar as inferring the former from the latter requires knowledge of the laws of nature. The predicates that designate the physical properties of the brain do not have a conjunctive structure such that part of the conjunction corresponds to a mental predicate. We cannot infer a priori, therefore, the possession of a mental property from the possession of a physical property.

3. Mental Causation and Downward Causation

The approach to the determination of macroscopic properties, based on laws of composition, that I have developed in this book solves two important problems. The first is what Kim calls the problem of causal exclusion. The second

35 See Armstrong (1997, Chapter 4.13). Worley (1997) has developed an analysis of the relationship between determinate and determinable universals similar to Armstrong's.

is to answer the question of whether the emergentist concept of downward causation is compatible with physicalism.

Let us start with the second problem. The view that macroscopic properties are determined by non-causal laws of composition provides the means for settling the debate between those who accept downward causation and those who take its possibility to be refuted by its incompatibility with well-entrenched metaphysical principles. Some emergentists[36] defend the thesis that a macroproperty can exert a constraint superposed[37] on the constraints exerted by the properties of the microscopic components; in this way, the macroproperty prevents the microproperties from determining the evolution of the system on their own. Their opponents[38] seek to show that downward causation is a myth incompatible with the metaphysical principle of "the causal closure of the physical domain" (Kim 2005, 15).

One of the targets of these critics is the concept of downward causation put forward by the neurophysiologist R.W. Sperry. He is an easy prey for defenders of microdeterminism and opponents of the possibility of downward causation because his position combines the fundamental thesis of the existence of macrodetermination with more controversial theses.

First, Sperry seems to take the structural form of a whole to be one of its *components*, in the same way as its matter. By attributing causal power to this form, he seems to return to a theory "in terms of components" as criticized by Broad (see Chapter 4). "I have repeatedly stressed the important causal role of the non-material space-time, pattern, or form factors and suggested that it is helpful to view any entity as . . . built of space-time components as well as of matter" (Sperry 1986, 266).

Second, the set of emergent properties to which Sperry attributes causal efficacy is vast and varied. It includes cases of macrophysical causality — "drops of water are carried along by a local eddy in a stream. . . . [T]he molecules and atoms of a wheel are carried along when it rolls down hill" (1969, 534) — informational causality — "computer software programs exert downward causal control over their electronic . . . correlates" (1986, 269)

36 In particular, Sperry (1969, 1976, 1986, 1992); Campbell (1974); Popper (1977); see also Gillett (2016).

37 Or that replaces it, in the case of the global properties of entangled systems of quantum physics.

38 See in particular Klee (1984); Kim (1992b, 1993c, 1998, 2005); Schröder (1998).

— biological causality — "the holistic properties of the organism have causal effects that determine the course and fate of its constituent cells and molecules" (1969, 533) — and psychological causality — "the subjective mental phenomena . . . influence and . . . govern the flow of nerve impulse traffic" (1969, 534). Even social properties can exert downward causal influence. Sperry describes them as "emergent forces of higher and higher levels that in our own biosphere include vital, mental, political, religious, and other social forces of civilisation" (1986, 269).

Critics of downward causation argue that it is incompatible with a number of principles whose truth is presupposed by scientific method. According to an argument put forward by Klee (1984), which I will consider in a moment, downward causation is incompatible with the universal truth of microdeterminism as well as with the principle that all causal relations are exercised through a mechanism (see also Craver and Bechtel 2007). Others, notably Kim, argue that downward causation is incompatible with the principle that there is only one independent complete causal explanation for a given event as well as with the principle of the causal closure of the physical domain. I will come back to this later. We will see that such arguments, even if they seem to be convincing with respect to Sperry's position, do not refute the theory of macrocausation developed here. In general, these arguments fail because they neglect the crucial distinction between causal and non-causal determination.

3.1. MACROCAUSATION WITHOUT AN UNDERLYING MICROSCOPIC MECHANISM

According to Klee, downward causal determination is intelligible only insofar as it is exercised through a mechanism. Advocates of downward causation could give us reasons to doubt that "micro-determinism" suffers no exceptions only "if a genuinely plausible mechanism of macro-determination could be provided" (Klee 1984, 61).[39] This reasoning suffers from the lack of a clear distinction between causal determination and non-causal simultaneous determination of the properties of an object by the properties of its parts. The conclusion that macrodetermination does not exist can be avoided by argu-

39 "We really have no established model of what a macro-determinative connection would be like. Direct determination from higher-levels to lower-levels seems somewhat mysterious when one attempts to construct a relatively precise scenario of the 'how' and the 'why' of it" (Klee 1984, 60).

ing that many types of causal determination are due to macroproperties that themselves are determined, in a non-causal way, by microproperties. Before returning to this model at greater length, I would like to offer, as an example of a causally efficacious macroproperty, the ability of hemoglobin molecules to bind oxygen (see Chapter 2.5; Rosenberg 1985, Chapter 4; Feltz 1995). It is the overall structure of the molecule that possesses this capacity. The overall structure is determined by non-causal laws of composition, grounded in the microscopic properties of the components of the hemoglobin molecule and their interactions. However, what is directly responsible causally for oxygen uptake is not the microscopic properties of the molecule but its overall structure. The function of hemoglobin is to carry oxygen from the lungs to the various tissues of the body. However, the property of the molecule that enables it to perform this function is a systemic property of the molecule that is "macroscopic" in a relative sense: it belongs to the whole molecule rather than to its constituents (i.e., its subunits and atoms).[40]

The macroscopic property of the molecule causally responsible for the fact that the molecule tends to bind an oxygen molecule in the lungs, and to release this molecule in the tissues, is its structure (i.e., its "conformation"). This macroscopic conformation can be generated, or determined, by a large number of configurations at the level of the constituent atoms. In a sense, there are many different kinds of hemoglobin in different biological species, each one different from the others at the atomic level. Hemoglobin is a complex molecule — a "tetramer" — made up of four chains of amino acids. Each type of hemoglobin is characterized by its own sequence of amino acids: this sequence is known as its "primary structure." However, hemoglobins differ from each other in most, but not all, of their 140 constituent amino acids: there are nine amino acids that occupy the same position in the primary structure of all hemoglobin molecules. The interactions between these nine component molecules are sufficient for a sequence to determine the specific conformation common to all hemoglobin molecules, which is then responsible for oxygen uptake. Insofar as there is only one common macroscopic conformation, it is also correct to say that there is only one type of hemoglobin molecule. In this case, "hemoglobin" is identified by its macroscopic property of having the conformation responsible for binding with oxygen.

40 Hemoglobin and its properties are macroscopic compared with the properties of its constituents, which are therefore relatively microscopic.

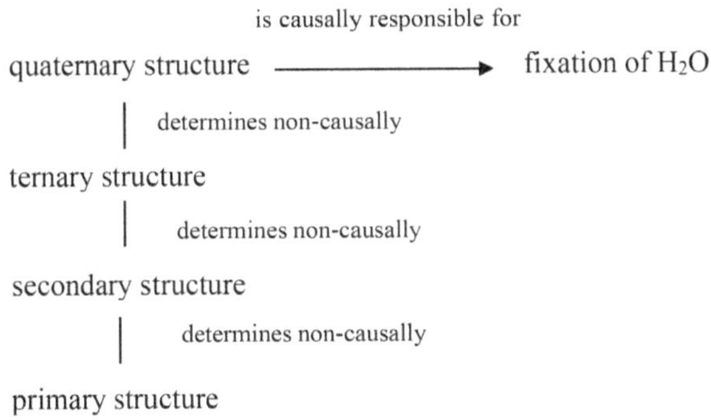

Figure 5.1 Non-causal determination and causal responsibility in hemoglobin.

The natural determination of the conformation of the whole molecule by the primary structure of the chain of amino acids — sketched in Figure 5.1 — involves two intermediate stages: interactions between the amino acids determine where the chain bends and folds, giving rise to the secondary structure. That structure brings together certain amino acids that would have been far apart in the primary structure, giving rise to new interactions that in turn determine the tertiary structure, the shape that the chain takes in space. Finally, hemoglobin is not, strictly speaking, *a* molecule but an aggregate of four molecules known as its subunits. Given the tertiary structure of the four subunits, they adopt a stable position in relation to each other, constituting the quaternary structure of the molecule: that is, of the aggregate, the oxygen-binding functional unit. This overall structure or conformation of the molecule, a systemic property of the molecule as a whole, is directly responsible causally for its interaction with oxygen: it is the conformation of hemoglobin that, in the relatively low pH environment of the lungs, causes oxygen molecules to bind to the iron molecules contained in the heme groups surrounded by each of the subunits.

The hemoglobin molecule illustrates the possibility that a macroscopic property has a causal responsibility of its own, in the sense that it is not an epiphenomenon of some underlying microscopic causation. It is not the properties of the atoms or amino acids making up hemoglobin that bind oxygen. Nor is this effect explicable on the model of the superposition of independent effects of each atom or each amino acid. None of these microscopic components has the tendency to bind to an oxygen molecule. Microscopic tendencies do not add up to a sufficiently strong attraction. On the contrary, this is a genuine case of macroscopic causality since the tendency to bind an oxygen molecule appears only at the level of the quaternary structure. There are no causal relationships at the level of the components of hemoglobin, the cumulative result of which would be to bind oxygen.

The case of the hemoglobin molecule shows that there is nothing mysterious about a causal relationship at the level of the whole system. On the contrary, on the basis of "what molecular biology has discovered about the hemoglobin molecule," chemistry provides "a direct and beautiful explanation of why the blood does what it does" (Rosenberg 1985, 74). Contrary to Klee's claim, the fact that a relationship of causal responsibility must be understood at a systemic — and thus (relatively) macroscopic — level does not entail that the "how" and "why" of its efficacy remain mysterious. The fact that the causally efficacious property is macroscopic does not imply that it is irreducible. On the contrary, the explanation of the relations that determine the causally efficacious property (the quaternary structure) by the primary structure, with its two intermediate stages, is a paradigmatic example of a successful microreduction.

We do not need to choose, as Klee's argument suggests, between macroscopic but mysterious determination and determination by a microscopic mechanism. According to my analysis, the mystery is unravelled thanks to the reduction of the efficacious property. This reduction involves the discovery and detailed description of non-causal determination relationships among primary, secondary, tertiary, and quaternary structures. However, the existence of the reductive explanation does not prevent the relationship of causal responsibility from involving an emergent property.

3.2. KIM'S ARGUMENT AGAINST MENTAL CAUSATION: PRELIMINARIES

Jaegwon Kim has presented an influential argument against the thesis that mental properties themselves can be directly causally efficacious. His argument has the form of a reductio of the hypothesis of mental causation. He begins by considering the hypothesis that one mental event causes another. Take the previous example, in which my thought that the noise in the street is disturbing my concentration caused my decision to close the window.

Schematically, the event in which I think that the noise disturbs me is represented by my possession of mental property M and the decision to close the window by my possession of mental property M^*. Kim's argument consists of showing first that M can have only a causal influence on M^* through an influence on the physical properties P^* underlying M^*. In other words, Kim seeks to show that mental causation, if it were possible, would presuppose downward causation. Then he argues that downward causation is incompatible with two principles widely accepted in metaphysics and epistemology, namely the principle of causal-explanatory exclusion and the principle of causal closure of the physical domain. According to the first principle, which is epistemological, "there can be no more than a single complete and independent explanation of any one event" (Kim 1988a, 233).[41] According to the second principle, which is metaphysical, "if you pick any physical event and trace its causal ancestry or posterity, that will never take you outside the physical domain. That is, no causal chain will cross the boundary between the physical and the nonphysical" (Kim 1997b, 282). This last principle directly denies the possibility that non-physical causes, particularly mental ones, can exert causal influences on physical events. However, we will see that the argument does not simply beg the question, insofar as Kim tries to justify this principle.

The debate about whether mental properties can be efficacious in influencing other mental or physical properties takes place against the background of a consensus on a physicalist conception of mental properties, which are instantiated locally in persons or animals endowed with cognition; they are intrinsic systemic properties of these individuals.[42] These systemic properties are determined exclusively by the physical properties of the individuals' parts and their interactions. This presupposes the strong supervenience of mental properties on the properties of the parts of the organism. Strong supervenience can be defined as follows.

---

41 See also Kim (1989a, 1989b). There is a similar debate about whether the teleological explanation of a given event is compatible with its mechanistic explanation or whether such explanations are mutually exclusive. Against Malcolm (1968), who argues for the thesis that they exclude each other, Heil (1992, Chapter 4) tries to show that these explanations are compatible insofar as they describe the same causal processes with different concepts and with different granularity.

42 It might be necessary to include part of the environment in the physical base of mental properties. This does not prevent this base from being located around the person. For reasons for including the environment, see Clark and Chalmers (1998); O'Regan and Noë (2001); Clark (2008).

> Necessarily, if a macroscopic object with parts $p_1, \ldots, p_n$ has a global property G, then there exist properties $P_1(p_1), \ldots, P_n(p_n)$ of the parts and relations between the parts $R_1(p_1 \ldots p_n) \ldots$ such that, necessarily, any macroscopic object that has parts $p_1 \ldots p_n$, with properties $P_1(p_1), \ldots, P_n(a_n)$ and relations $R_1(p_1 \ldots p_n), \ldots$ has the property G.

In Chapter 4, I explicitly required that emergent properties satisfy this physicalist condition of mereological determination. This conception of emergence is consistent with Kim's thesis that "a particularly important and promising approach . . . is to explicate mind-body supervenience as an instance of mereological supervenience. That is, we try to view mental properties as macroproperties of persons, or whole organisms, which are determined by, and depend on, the character and organization of the appropriate parts, or subsystems, of organisms" (1993b, 168; see also Kim 1978, 1988b, 1990).

Kim judges correctly that the problem of mental causation is a special case of the more general problem of understanding when and how different properties of an event cooperate in their causal influence on other events and when, on the contrary, the causal efficacy of one property excludes the causal efficacy of another. The dispute concerns the question of whether a mental property *M* can have its own causal efficacy in bringing about another mental property *M**, or whether the impression of such causal efficacy is illusory, since the only relationship of causal efficacy relates the underlying physical properties *P* and *P**. We can sketch the situation as in Figure 5.2.

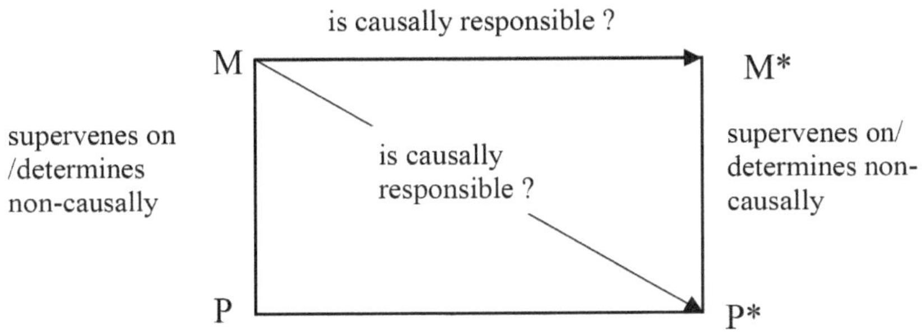

Figure 5.2 The controversial causal responsibility of mental properties.

Before considering in detail Kim's argument to the effect that only physical properties can be causally efficacious, it is useful to mention three terminological points. The first concerns the ontology of the terms of causal relation: according to Kim, events linked by causal relations are structured entities composed of an object, a property that the object exemplifies, and an instant at which the object exemplifies the property. The effect considered in our example consists for Kim of a triplet <*s*, *M*, *t*>, where *s* represents the bearer of the property (i.e., the person who decides to close the window), *M* the property of deciding to close the window, and *t* the instant at which this decision is made. However, in the context of his analysis of mental causation, Kim expresses himself more simply by talking about *the property* that causes or determines another property. This does not create any misunderstanding insofar as the context fixes the object and the instant univocally.

In this simplified terminology, Kim's thesis is this: the fact that *M* causes *M** is only an alternative way of conceiving of the only real causal relationship, namely that between *P* and *P**. Instead of saying that the event <*s*, *P*, *t*> causes the event <*s*, *P**, *t**>, we can simply say, with Kim, that *P* causes *P**. There should be no misunderstanding about the relevant conception of properties: when we ask whether a "mental property *M* is causally efficacious with respect to physical property *P**" (Kim 1993c, 207), it is clear that local instances of these properties, not universal properties, are at stake. From now on, it will be understood that this means "that a given instance of *M* causes a given instance of *P**" (Kim 1993c, 207). Similarly, given that we are dealing with causation and that the concept of causation requires the temporal anteriority of the cause with respect to the effect, properties of effects — designated by starred predicates (here *P**) — are systematically exemplified later — namely at *t** — than properties without a star (here *M*) — namely at *t*. A statement that one property *M* causes another *M** can always be translated into the language of causation between *events*: in the conception of events as concrete particulars, the mental property *M* and the physical property *P* on which it supervenes and by which it is determined belong to the same concrete event.[43] The question of the respective roles played by *M* and *P* is posed in this context in terms of causal efficacy or causal responsibility. Which fact is causally responsible for the fact that the effect-event has the property *P**: the fact that the cause-event has *M* or the fact that it has *P*? Or, equivalently, which of the properties *M* and *P* of the cause is efficacious in ensuring that the effect has the property *P**?

Another terminological detail concerns the object to which the properties are attributed. If *M* belongs to a subject, then we can consider the question of its efficacy in relation to the underlying physical properties, insofar as they belong to the subject's parts, particularly neuronal parts, or insofar as they belong to the same subject as a whole. In order to express himself in the second way, Kim uses the concept of a *micro-based property* (see Chapter 4.8): "*P* is a *micro-based property* just in case *P* is the property of being completely decomposable into nonoverlapping proper parts $a_1, a_2, \ldots, a_n$, such that $P_1(a_1)$, $P_2(a_2), \ldots, P_n(a_n)$, and $R(a_1, \ldots, a_n)$" (Kim 1998, 84). This is a technical means of attributing the physical properties of the parts of an object to that object as a whole, without prejudging whether the property thus logically constructed and attributed to the object is a real property or only a nominal one: a real property has causal powers of its own, whereas many merely nominal properties do not.[44] The table that I am sitting at, for example, has the disjunctive property of being rectangular or yellow, but this property is not "natural" or "real"[45] insofar as it does not give the table any causal power of its own, different from the powers due to the properties of being rectangular and of being yellow.

43 We must be careful to avoid a terminological confusion. What Kim calls an "event" is not a concrete particular *c* occupying space-time, which has many properties, but what we call a "fact": the fact that event *c* has property *P* at time *t*. See Kistler (1999a, 1999b, 2006d).

44 Lewis (1983) calls these properties "abundant."

45 Lewis (1983) calls them "sparse" properties.

A third terminological point should be clarified before I examine Kim's argument. We must not confuse the physicalist thesis that I expressed above as the thesis of mereological supervenience with the "Physical Realisation Thesis" (Kim 1993b, 344; 1993c, 198). According to the latter thesis, "all mental properties are physically realized; that is, whenever an organism or system instantiates a mental property *M*, it has some physical property *P* such that *P* realizes *M* in organisms of its kind" (1993b, 344; 1993c, 198). The first thesis concerns the micro-macro relationship among the parts of an organism and its global properties; the second concerns the relationship between second-order functional properties and the first-order properties that perform the function. For example, to conceive of the property of experiencing pain as a functional property is to conceive of it as a second-order property: the property of having a first-order property from within a set *B*. The properties in *B* satisfy the functional constraints specific to pain: being caused by damage to the body, causing behaviour aimed at avoiding the cause of that damage, causing the desire for the pain to cease, and so on. According to the physical realization thesis, mental properties are actually physical properties conceived of by second-order concepts (Kim 1998, 104). To have a mental property is not, according to this conception, to have a first-order property (i.e., a causally efficacious property). Rather, it means having a second-order property: the property of having a physical property that obeys certain causal or functional constraints. I am not denying that it is possible or appropriate to conceive of mental properties in this functional way. I am only contesting the legitimacy of inferring from this possibility that mental properties are not first-order properties. The fact that a property can be described with a second-order predicate does not prove that it is impossible to describe the same property with a first-order predicate. Thus, it remains possible, as emergentism envisages, that mental properties are macroproperties that "can, and in general do, have their own causal powers, powers that go beyond the causal powers of their microconstituents" (Kim 1998, 85). In other words, properties that fulfill the functional roles corresponding to second-order functional concepts can be macroscopic mental properties that can also be designated by first-order predicates (see Chapter 3).

Here is what is at stake in the debate: I defend the thesis that an emergent mental property is a first-order property, determined by the physical properties of the brain's components according to laws of composition, which possesses causal powers not possessed by the physical properties that determine its existence. Against this, Kim argues that the only causally efficacious properties are physical or physiological properties (i.e., properties that belong to the parts of the person's body), although the concept of a micro-based property makes it possible to attribute them formally to the person herself.

3.3. THE FIRST PART OF KIM'S ARGUMENT: NO MENTAL CAUSATION WITHOUT DOWNWARD CAUSATION

As I said earlier, Kim's argument proceeds in two stages. First, Kim explains that the only way M could cause M^* is by downward causation; M can cause M^* only by causing P^*, the physical property underlying M^* by virtue of mereological supervenience. Second, he tries to show that such a downward causal influence is incompatible with the principles of causal-explanatory exclusion and of the causal closure of the physical domain.

Here is the first part of the argument. Kim says that, for a given mental property M^* (as sketched in Figure 5.2), there are two possible answers to the question of what is responsible for M^*. According to the first, M^* is due to M; according to the second, M^* is due to P^*. Kim then argues that there are only three ways of reconciling these two answers. Of the three, he tries to show that the first two are unacceptable. The third is therefore provisionally retained, although the second part of the argument shows that it too must be abandoned, depriving M of all causal power both to cause P^* and to cause M^*. The possible answers are

(1) M and P^* are jointly responsible for M^*;

(2) M and P^* are each responsible for M^* so that M^* is overdetermined; and

(3) M causes M^* by causing P^*.

According to hypothesis (1), M and P^* taken together are responsible for M^*, although neither alone is sufficient for M^*. Kim rejects this hypothesis on the ground that the supervenience thesis implies that P^* alone is sufficient for M^*. He argues against assumption (2) — that M and P^* overdetermine M^*, in the sense that each alone is sufficient for M^* — by proposing that the concept of overdetermination would be appropriate in the present case only if M^* had "two distinct and independent origins in M and P^*" (1993c, 205). However, as the situation is conceived of, M and P^* are not independent. Compare the situation with the paradigmatic case of independent overdetermination described above: when two soldiers in a firing squad both fire a bullet at the victim, such that each is sufficient in itself to cause her death, each causes the victim's death. Since we are not dealing here with such an exceptional situation in which two independent causal chains converge on the same effect, Kim applies the principle of explanatory exclusion: there is only one complete and independent explanation for M^*. From the fact that P^* alone is sufficient for M^*, he concludes that M is not a complete cause of M^*, in the sense of constituting a complete causal explanation of M^*.

This argument invites the objection that it fails to make the crucial distinction between causal and non-causal determination; this neglect goes hand in hand with the neglect of the difference between the relationships expressed by "determines," "is a sufficient condition for," and "is causally responsible for."[46] If we distinguish these concepts of determination, then it appears that hypotheses (1) to (3) do not exhaust all possibilities. There is a fourth way of conceiving of the relationships of determination among M, P^*, and M^* that does justice to the premises of Kim's argument while avoiding its conclusion.

The argument against hypothesis (2) shows that what Kim means by "overdetermination" is independent causal overdetermination such as it exists in the case of the firing squad, in which several independent causal processes lead to the victim's death. However, M and M^* are indeed in a relationship of causal responsibility, but the determination between P^* and M^* is not causal:

---

46 Thomasson (1998) and Jacob (2002) express similar criticisms. Slors (1998) offers a more charitable interpretation of Kim's argument. He asks how we can conceive of the relationships of realization (between P^* and M^*) and causation (between M and M^*) in such a way as to make them sufficiently "similar" to make Kim's claim intelligible that M and P^* are in competition for being a "sufficient condition" for M^* and to make intelligible the solution that Kim considers (this is hypothesis (3) mentioned in the text), in which these relations are linked transitively, such that one is a sufficient condition for the next: $M — P^* — M^*$. According to Slors, the only possibility is to interpret both realization and causality as nomic relationships. Given the traditional interpretation of the concepts of realization and causality, this indeed seems to be their "greatest common denominator" (i.e., the richest conceptual element common to both). Slors accepts the idea that, in this interpretation of the two relationships, M and P^* compete to be nomically sufficient for M^*. He concludes that there are only two ways of justifying mental causation and resisting the verdict that the mind is merely an epiphenomenon: one must deny that either causation or realization is nomic. However, as I show in the text, these relationships of nomic sufficiency, belonging to very different species of relationships, are not in competition; therefore, it is enough to distinguish them to solve the problem raised by Kim. There is no need to resort to the radical solution proposed by Slors to deny the nomic character of realization.

it is the relationship between a neuronal property and the mental property that emerges from it as a function of a law of composition.[47] First, an emergent property and the properties from which it emerges belong to the same object at the same time, whereas a causal relationship requires its terms not to overlap in space-time.[48] Second, emergence is based on nomic necessity, whereas causation is contingent. Since P^* and M^* are not in a causal relationship, M and P^* cannot overdetermine M^* in the sense of causal overdetermination.

In other words, both M and P^* determine M^*, though not in the same way: M causes M^*, by virtue of a nomological dependence between these properties, whereas P^* determines M^* in a non-causal way. Thus, there is a fourth way, neglected by Kim, to reconcile the two explanations of the presence of M^*: the first is causal, whereas the second is a non-causal explanation (which we can call compositional). M and P^* do not causally overdetermine M^*, although each alone is sufficient for M^*. In other words, the expression "is sufficient for" can have two meanings: it can express the relationship of causal responsibility and the relationship of non-causal compositional determination.

Here are two simple examples showing that explanations belonging to these two categories answer independent questions so that they complement rather than exclude each other. Why is there an equilateral triangle on the paper on my desk? First, the causal answer: because I drew it. Second, the non-causal answer: because there is an equiangular triangle on the paper, and all equiangular triangles are necessarily equilateral. Why is this gas at a temperature of $T = 50°\ C$? First, the causal answer: because I have just increased its temperature by heating its container. Second, the non-causal answer: because its pressure and volume are P and V, because it is an approximately ideal gas, and because T is proportional to the product of P and V, by virtue of the ideal gas law.

47 Crane (1995, 232) and Loewer (2002), among others, point out that the thesis that we are dealing with overdetermination lacks plausibility only insofar as it is interpreted on the model of independent overdetermination of the firing squad. It is not plausible that effects of mental causes are also *independently* caused by the physical properties underlying those mental properties. This leaves open the possibility that mental causation involves systematic overdetermination by mental and physical properties linked by psychophysical laws.

48 Kim (1998, 44) accepts the Humean requirement that cause and effect must be separated in time.

The existence of two determinations of the same property instance, each "complete" in its own way and independent of the other, is certainly incompatible with the principle of explanatory exclusion, according to which "there can be no more than a single *complete* and *independent* explanation of any one event" (Kim 1988a, 233). In this context, "complete" means "in itself sufficient," and "independent" means "without conceptual link or link of logical or metaphysical necessity." However, the coexistence of our two determinations is compatible with another principle, which can easily be confused with the principle of explanatory exclusion. According to the principle of *causal-explanatory* exclusion, it is exceptional for there to be two complete and independent *causal* explanations of an event. Such a principle simply does not apply in the cases that we have considered, insofar as one of the two determination relationships is non-causal.

We therefore arrive at the following analysis of our schematic situation of mental causation. Let us assume that there is a psychological law linking instances of M to subsequent instances of M^*. The instance of M^* can then be *causally* explained by the preceding instance of M, together with the psychological law in question. This does not prevent the possibility that the same instance of M^* can be explained alternatively non-causally, in terms of the physical or physiological properties P^*, on the basis of the relevant law of composition by virtue of which P^* gives rise to M^*.

If this analysis is correct, then I have refuted Kim's argument that the only way to account for the fact that M^* is determined both by M and by P^* is to assume that M causes M^* *by causing* P^* (i.e., by virtue of downward causation).

Furthermore, we have seen that a "principle of explanatory exclusion" is not plausible. There is no reason to deny the possibility of two explanations of the same fact when these explanations belong to *different categories* — one being causal and the other non-causal. This is true even if each can be considered complete in its own right, in the sense that it allows us to deduce the *explanandum*.[49]

In defence of Kim's argument, it could be argued that the principle of explanatory exclusion should be interpreted as a principle of *causal-explanatory* exclusion. According to this principle, there cannot be more than one

49 Kim (2002, 2005) challenges this result, arguing that there is a tension between the non-causal explanation of M^* by P^* and the causal explanation of M^* by M.

complete and independent *causal* explanation for a given event. After all, Kim defends an "explanatory realism" according to which any true explanation is based on a *causal* relationship between the terms of the explanation: "To 'have an explanation' of event *e* in terms of event *c* is to know, or somehow represent, that *c* caused *e*; that is, . . . explanations of individual events are represented by singular causal propositions" (Kim 1988a, 230).

It is crucial here to bear in mind the distinction mentioned above between two ways of conceiving of the terms of causal relationships: either they are events in the sense of particulars that occupy a portion of space-time and have many properties, or they are facts — which Kim calls "events" — that is, propositional entities. In the first conception, the cause *c*, as a particular event, can possess both the physical property *P* and the mental property *M* and the effect event *e* both the physical property *P** and the mental property *M**. There is no reason, then, as Marras (1998) has observed, to deny the possibility that the fact that *c* is *M* is causally responsible for the fact that *e* is *M** and that, so to speak in parallel, the fact that *c* is *P* is causally responsible for the fact that *e* is *P**.

To evaluate the principle of causal-explanatory exclusion, we must distinguish between its application to events and its application to facts: it is doubtful whether it makes sense to speak of a complete explanation of a particular event. What we are trying to explain are always facts: insofar as an explanation always takes the form of an argument whose *explanandum* constitutes the conclusion, only a propositional entity can be the object of an explanation. However, when it is said that there is only one complete causal explanation of an event, this can mean two things. First, it can mean denying the possibility of downward causation. It denies that two different facts about the same event, in our case the fact that *c* is *P* and the fact that *c* is *M*, cause the same physical fact, that *e* is *P**. We will come back to this in a moment. Second, it can mean that there cannot be two causal relationships of responsibility concerning the same pair of events: on the one hand, the fact that *c* is *P* is causally responsible for the fact that *e* is *P**; on the other hand, the fact that *c* is *M* is causally responsible for the fact that *e* is *M**. The principle of causal-explanatory exclusion does not seem to apply to the latter case: the two relations of causal responsibility do not seem to exclude each other. This is not a case of causal overdetermination since their effects are different. Moreover, the existence of two parallel explanations at different levels can be explained by the fact that their truth makers are not metaphysically independent. Since, according to physicalism, for every mental property M there is an underlying physical property that determines it, mental properties depend on physical properties, and mental explanations depend on mental properties. If there is a physical explanation, however, the psychological explanation depends on this physical explanation. However, as we will see, the dependence of mental properties on physical properties does not exclude the possibility that there is no physical explanation for a pair of causally related events, in which case the only accessible explanation is psychological.[50]

As we saw in Chapter 1, certain elementary mechanisms of learning have been reduced to neurophysiology, in the sense that microscopic mechanisms have been discovered that give rise to the regularities observed at the psychological level. As far as classical (i.e., Pavlovian) conditioning is concerned, Rescorla and Wagner (1972) discovered that the reinforcement of the associative strength between an unconditional stimulus (US) and a conditional stimulus (CS) X (see Chapter 1.9), in an experimental session in which X is presented just before the US, obeys a law expressed by the following formula: $\Delta V_X^n = \alpha_X \beta(\lambda - V_{AX}^{n-1})$.

This formula corresponds to the general case in which the same US has already been used to condition the subject to another conditional stimulus, A. ΔV_X^n represents the increase in associative strength with which the CS X triggers the response (naturally appropriate for the US), as obtained at

50 Marras is certainly wrong when he says that, on the contrary, "each of the two explanations appears to be complete in its own domain of application, and each appears to be independent of the other (entailing, as they do, distinct counterfactuals)" (1998, 449). The disagreement stems from the fact that Marras accepts the Davidsonian thesis of the irreducibility of psychological explanations to neurophysiological explanations. Within this framework, Marras infers from the fact that $M \rightarrow M^*$ (the fact that c is M is causally responsible for the fact that e is M^*) implies the truth of the counterfactual $\neg M \Box\!\rightarrow \neg M^*$, and from the fact that $P \rightarrow P^*$ implies the *different* counterfactual $\neg P \Box\!\rightarrow \neg P^*$, that $M \rightarrow M^*$ is independent of $P \rightarrow P^*$. As soon as we abandon the doctrine of irreducibility, the argument loses its validity: both implications can be correct, whereas $M \rightarrow M^*$ is dependent on $P \rightarrow P^*$. If M depends on P and M^* depends on P^*, then it is possible that the whole process leading from the instance of M to the instance of M^* depends on a parallel underlying causal process leading from the instance of P to the instance of P^*. (For downward causation, this scenario is considered by Bennett [2003] and Witmer [2003]. I come back to this later in the text.) To refute Marras's reasoning, it is sufficient to show that this scenario is possible. It remains an open question whether any relationship of psychological causal responsibility is indeed accompanied by an underlying relationship of physical causal responsibility. I would suggest that, given the complexity of the neurophysiological and even more so the microphysical details, we can expect this not to be the case.

nth exposure; α_X represents the salience of the stimulus X, β the salience of the US, and λ the maximum associative strength that can be obtained by association with the US; finally, V_{AX}^{n-1} represents the total associative strength reached between the US and the two CSs, X and A, in the previous n–1 exposures. This total strength V_{AX}^{n-1} is the sum of the individual association strengths of the two conditional stimuli A and X, so that $V_{AX}^{n-1} = V_A^{n-1} + V_X^{n-1}$.

This law can form the basis of a relationship of causal responsibility between one mental property and another and thus illustrate the situation outlined in Figure 5.2. The earlier mental state M contains the associations between A, X, and the US created during the first n–1 exposures as well as the experience of X preceding the US in the nth exposure; the law predicts that this produces a new mental state M^* that contains associations whose strength is a function of M, as expressed by the law. This law applies to various animal species (Rescorla and Wagner report the results of experiments with rabbits and rats); this makes it unlikely that there is a single underlying microscopic property common to different species for each mental state that emerges during learning.

I have shown that the first step of the argument by which Kim tries to establish the impossibility of mental causation is not conclusive. This does not diminish, however, the importance of the second step of his argument. In that step, Kim intends to show that the hypothesis of the existence of downward causal relations is incompatible with the principle of the causal closure of the physical domain as well as with the principle of causal-explanatory exclusion. Even if such downward causation does not play a role in every causal relationship between mental events, the emergentist position defended in this book in fact presupposes that emergent properties, such as cognitive properties, can have a causal influence on events situated at lower levels of complexity, in particular on physiological events: my decision to close the window is an emergent property of my person, which leads causally to the modification of the state of the motor neurons involved in the execution of the appropriate action.

3.4. THE SECOND PART OF KIM'S ARGUMENT: NO DOWNWARD CAUSATION

In Figure 5.2, the downward causal influence is represented by the diagonal arrow: it corresponds to the hypothesis of a causal relationship between the thought that the noise bothers me (M) and the microscopic neuronal events

(P^*) that are part of the basis that determines the decision to close the window (M^*). To demonstrate the incompatibility of this hypothesis with the principles of causal closure of the physical domain and causal-explanatory exclusion, Kim begins by arguing that there are four possible causes of P^*.[51]

(1) *M* and *P* together constitute a sufficient cause of P^*;

(2) *M* and *P* are two mutually distinct sufficient causes of P^*, so that *M* and *P* overdetermine P^*;

(3) *P* causes P^* through *M*; and

(4) *P* causes P^* directly, without any causal contribution from *M*.

Only the first three hypotheses presuppose the existence of downward causation. If there are no other possibilities, then Kim only needs to refute these three scenarios in order to refute the possibility of downward causation.

3.4.1. The Refutation of Scenario (1)

Against hypothesis (1), Kim (1993c, 207) argues that the physical realization thesis implies that *P* by itself is sufficient for P^*. Moreover, if *P* is sufficient for *M*, and if *P* and *M* together are sufficient for P^*, then *P* alone is sufficient for P^*. As I pointed out earlier, the choice of conceiving of *M* as a second-order functional property and *P* as the first-order property that realizes it would prejudge the question of the causal efficacy of *M* in the negative. By its very conception, a second-order property has no causal efficacy of its own because causal efficacy is limited to the level of the first-order properties that realize it. However, we can express an argument similar to Kim's that respects our conceptual framework: if *M* is a first-order macroscopic property, and if *P* is the set of microscopic properties that determines it according to a law of composition, then *P* is sufficient for *M*. However, in the latter case, this is a sufficient condition for nomological reasons and not for logical or conceptual reasons,

51 The expression "cause of P^*" is ambiguous: it can have the meaning of causality between events and the meaning of causal responsibility between facts. However, given Kim's terms of the causal relationship, "is the cause of" is always equivalent to what I call "is causally responsible for."

whereas Kim's functionalist conception takes the relationship between P and M to be the conceptual relationship of realization.

We can interpret (1) in two ways, corresponding to two ways in which P and M can "join forces," in the sense of each contributing to the effect P^*. In both interpretations, P alone is sufficient for M. The relationship of being a sufficient condition here covers both the relationship of non-causal determination — between P and M — and the relationship of causal determination. The difference between the two interpretations concerns the latter: according to the first interpretation, M alone is causally responsible for P^* by virtue of a causal law. In other words, P^* exists only because — and insofar as — it is determined by M, but P does not take part in the direct causal responsibility for the production of P^*. We can then say (although this is true only *ceteris paribus*, as with all applications of the laws of the special sciences) that M is "sufficient for P^* in the circumstances."

According to the second interpretation,[52] both P and M exert a constraint on the evolution of the system, but neither determines the state of the system, at the time of the event P^*/M^*, completely on its own. M determines only a certain framework within which the system must evolve but not its detailed evolution. This constraint can be construed by analogy with the constraints exerted on a mechanical system. The rail line on which a train is travelling does not in itself determine the trajectory of the train: the forces acting on it — notably gravitation, the force released by the locomotive engine, and the force of friction — determine the direction and speed with which it moves. However, the rail line determines a framework that restricts the number of degrees of freedom available to its movements to two: direction and speed. In the case of a ball rolling inside a bowl, the internal surface of the bowl determines a frame that limits the possible trajectories of the ball. This surface exerts a constraint on the trajectory of the ball that leaves it with only four degrees of freedom or four dimensions in what is known as the phase space of its trajectory: the two dimensions of the surface and the two dimensions of the velocity. Without any constraint, such as that exerted by the rail line or the surface of the bowl, the displacements can occupy all six dimensions of the phase space: three dimensions for the position in space and three dimensions for the components of the velocity in the three spatial dimensions.

52 The concept of downward causation is developed in Kistler (2009, 2017, 2021).

To return to the determination of the physical state P^* of a cognitive system, the second interpretation of hypothesis (1) amounts to supposing that the mental state M defines a framework that limits the possibilities of evolution of the system without determining them completely in detail. Let us assume that there is a psychological law $M \rightarrow M^*$ stating that, *ceteris paribus*, any system in M evolves toward state M^*. At the instant of M, the system has a well-defined physical state P that determines M in a non-causal way. But M is compatible with a certain number of other physical states that would determine a mental state of the same type as M. In other words, M does not determine P in detail, only a certain framework that delimits a space of possibilities. This thesis corresponds to the thesis of "multiple realizability" in the sense of micro-macro determination, which we examined in Chapter 2. Similarly, M^*, although causally determined by M by virtue of the law $M \rightarrow M^*$, does not determine the details of its physical realization, P^*. Rather, M^* constitutes a framework that constrains the evolution of the system, the details of which are determined at the physical level (i.e., by P).

By virtue of the transitivity of the relationship of being a sufficient condition, it is correct to say in both interpretations of hypothesis (1) that the instance of P is sufficient for the instance of P^*. Without distinguishing between the two interpretations, Kim rejects (1) because it lacks parsimony. However, his argument is based on a strong presupposition: following the (Davidsonian) principle of the nomological character of causality, Kim assumes that the causal relationship between M/P and M^*/P^* is determined by *physical* laws.

According to Kim,

> P appears to have at least as strong a claim as M as a direct cause of P^* (that is, without M as an intervening link). Is there any reason for invoking M as a cause of P^* at all? The question is not whether or not P should be considered a cause of P^*; on anyone's account, it should be. Rather, the question is whether M should be given a distinct causal role in this situation? I believe there are some persuasive reasons for refusing to do so. (1993c, 207)

Since there are physical laws that determine P^* on the basis of the earlier state P, the assumption that M also contributes to P^* — whether in the first or the second interpretation of (1) — violates the principle of explanatory simplicity, which finds its precise expression in the principle of causal-explanatory exclusion. The hypothesis that the complete physical cause P^* is systematically accompanied by a second cause M, itself complete as in the first interpretation of (1) or incomplete as in the second, must be excluded.

The question of the level of the laws that determine P^* is empirical and cannot be decided a priori, on mere conceptual grounds, as Kim's argument presupposes. This applies in particular to the hypothesis that, for any physical event P^* taking place at time t, there exists, at each preceding time t, a complete and purely physical cause P of P^*. This is an empirical hypothesis that has no more a priori credibility than the hypotheses underlying the two interpretations of (1): according to the first, there is a law by virtue of which M alone determines P^*; according to the second, there is a psychological law $M \rightarrow M^*$ by virtue of which M determines a framework that restricts the degrees of freedom of the evolution of the system and physical laws that determine, within the framework fixed by the psychological law, the details of the physical evolution from P to P^*.

No a priori argument can establish which of these three possibilities is correct. Therefore, Kim is wrong to assume that the first hypothesis, according to which P alone determines P^* by virtue of laws at the physical level, needs no empirical justification. Indeed, it is empirically possible that there is no law at the level of P and P^* that would make P^* predictable from P: this might be the case of a chaotic system in the physical sense. If P describes the state of a chaotic system in the basin of a strange attractor,[53] then for any accuracy or tolerated error there is a time in the future such that the states of the system after that time cannot be predicted with that accuracy. In such a situation, it is possible that (1) is correct, according to one of its two interpretations. I will come back to this later.

3.4.2. The Refutation of Scenario (2)

Against hypothesis (2), Kim takes the hypothesis that M and P causally overdetermine P^* to be "absurd" (1993c, 208). This judgment is plausible insofar as causal overdetermination is taken to mean the parallel actions of two *independent* causal processes that lead to the same effect, so that each would

53 This term was introduced in Chapter 4.8.

have been sufficient without the other for the effect to occur. The paradigmatic case is the firing squad, in which several bullets reach the heart of the victim by causally independent paths at the same instant. It is certainly not plausible for cerebral and mental causes to act systematically *independently of each other* to cause P^*. Independent but converging causal processes such as those occurring in the scenario of the firing squad are rare and exceptional. Kim expresses this by saying that the assumption that each mental cause is accompanied by an *independent* physical cause is incompatible with the principle of causal-explanatory exclusion, mentioned above. Two complete causal explanations of a fact create, he says, "an unstable situation requiring us to find an account of how the two purported causes are related to each other" (1998, 65). There are good reasons to abandon the hypothesis that, if a mental cause M brings about a physical effect P^*, then it always overdetermines P^*, in the sense that P^* always also has a parallel complete physical cause P.

It is less easy to show why the other interpretation of overdetermination also leads to an absurd consequence. It seems to be possible that mental causes M and their underlying properties P overdetermine their physical effects P^*, in the sense that each is sufficient for P^* by virtue of a causal law (purely physical in one case and psychophysical in the other) but without one being independent of the other.[54] Indeed, several authors have suggested that the efficacy of mental causes is a matter of "dependent overdetermination" (Witmer 2003, 205). In a situation in which the effect is overdetermined in this way, the fact that the mental property is efficacious in causing the effect depends on an underlying physical process. Once dependent overdetermination is clearly distinguished from "autonomous overdetermination," in which neither of the two causes depends on the other, it seems to be conceivable that the former provides an appropriate model for mental causation. Indeed, the only reason to deny that mental causes systematically overdetermine their effects presupposes that overdetermination is always autonomous. Indeed, a large-scale "systematic coincidence," without any basis in mutual dependence or in relation to a third factor, is certainly not plausible. Bennett (2003)

54 Kim (1998, 52–53) objects to Block (1990) for failing to distinguish between overdetermination by independent causes, not a plausible hypothesis in this case, and overdetermination by causes that are not independent. To exclude this possibility, Kim appeals to a "causal inheritance principle" (54), according to which the "two causes" in reality are a single cause conceived of in two ways, respectively by a first-order concept and by a second-order concept.

defends what she calls "compatibilism": that is, the thesis according to which the causal efficacy of a mental property M is not "excluded" or "pre-empted" but perfectly compatible with the causal efficacy of the underlying physical property P. She writes that, "if a mental cause is efficacious in bringing about some effect, the only physical causes that are also efficacious in bringing about that effect are ones that necessitate the mental cause" (2003, 487).[55] However, none of these authors considers the possibility that, in certain situations, the mental cause might be an indispensable component of what is causally responsible for some physical fact P^* concerning the subject's body at t^*. This would be the case if the evolution of the underlying neural state were chaotic in the sense of systems theory. In this case, there is a time t earlier than t^*, such that in principle there is no knowable neuronal state P at t, which could be used to explain P^* deductively. The mental cause would be necessary — possibly together with the neuronal state — to explain the bodily movement in question.

To draw a metaphysical conclusion — that the state P^* of all the parts of the system at t^* is not causally determined by the state P of all the parts of the system at t — from the impossibility of long-term prediction in a chaotic system, two presuppositions have to be made. The first concerns the interpretation of the notion of causal determination. As we saw earlier (Chapter 5.2), causal relationships can be analyzed at two levels.

> 1. Causal relationships can be conceived of as relationships between particular events, where "particular" is taken to mean a concrete object or event with many properties. We assume that causation is based on the transmission of a quantity of

55 According to Pereboom and Kornblith,

> the psychological explanation of an event does not compete with its physical counterpart because the mental causal powers referred to in the psychological explanation are wholly made up of the physical causal powers referred to in the physical explanation. Hence, the claim that a bit of behavior was caused by certain mental states is not an explanation which competes with the physical account which underlies it, any more than the claim that I secured ice-cream with cash competes with the claim that I secured ice-cream with bits of paper and metal. (1991, 143–44)

The analyses by Pereboom and Kornblith, Bennett (2003), and Witmer (2003) try to rescue the causal efficacy of mental properties by construing them as physical properties "otherwise conceived." I have analyzed this strategy, which is also that of Kim, Jackson, and Chalmers, in Chapter 2. I will come back to it later in this chapter.

energy (or some other conserved quantity) from one event to the other.

2. However, in most contexts in which we are interested in a causal relationship, in particular in the context of scientific explanation, we conceive of causation in terms of certain well-defined *properties* of events: for example, we do not just want to establish that *there is* a causal relationship between two successive episodes in the life of an organism learning an association by classical conditioning. Instead, the aim is to understand a *fact* F_2 about the organism at time t_3 after a learning episode, as a function of a *fact* F_1 about the learning episode, so that F_1 is causally responsible for F_2. F_2 might be the fact, for example, that the associative strength V_A between a conditioned stimulus (A) and an unconditioned stimulus (US) increased by ΔV_A during a conditioning episode in which the subject was exposed, at t_1, to a stimulus consisting of A and X, before being exposed, at t_2 immediately following t_1, to the US. According to Rescorla and Wagner (1972), what is causally responsible for F_2 is a fact that relates to the cognitive system at time t_1 and to the learning episode that takes place between t_1 and t_2: F_1 relates to the associative strengths V_A and V_X of stimuli A and X before the conditioning episode and various parameters. F_1 causally determines F_2 by means of a law, according to which the change in associative strength is equal to $\Delta V_A = \alpha_A \beta (\lambda - V_{AX})$, where λ is the maximal strength of association that can be obtained with the US, α_A and β are "learning rate parameters" (Rescorla and Wagner 1972, 76), α_A being specific for the A and β for the US, and V_{AX} is the combined associative strength of stimuli A and X at t_1, with $V_{AX} = V_A + V_X$. We can interpret the causal determination of F_2 by F_1 in an ontological way: the associative strength after the learning episode is independent of our knowledge about and descriptions of these facts. To interpret this second aspect of the causal relationship, we can start from the deductive-nomological analysis of causal explanation: to explain F_2 causally is to produce a deductive-nomological argument of which F_2 is the

conclusion and whose premises contain the initial condition F_1 as well as a certain number of statements of laws of nature. From a realist perspective, we can infer that, for this causal explanation to be true, there must exist facts F_1 and F_2 and laws *N* expressed by the nomological statements, such that F_1 is causally responsible for F_2, given the laws *N*.[56]

The second presupposition concerns the interpretation of indeterminacy in a chaotic system:[57] not only is it epistemic, but also it has an ontological interpretation. When it comes to determining the value of a measurable quantity in a physical system that takes its values in a continuum, we cannot attribute any empirical meaning to the hypothesis that this quantity has a value known with absolute precision. We might suppose that measurable quantities nevertheless objectively possess values of infinite precision. If we assume determinism, then the evolution of a system is objectively determined with infinite precision within any arbitrary time. However, the state of a physical system at time *t* can only be known, even in principle, with finite precision. If the system is chaotic, then for any finite precision or margin of error there is some time t^* in the future such that the state of the system after *t* cannot be determined with a precision within that margin. Therefore, if we consider states that can be known at least in principle, then such knowable-in-principle states of the system at *t* do not determine knowable-in-principle states of the system for times after t^*.

3.4.3. The Refutation of Scenario (3)

Against scenario (3), Kim offers three arguments. First, "given the simultaneity of the instances of *M* and *P* respectively, it is not possible to think of the *M*-instance as a temporally intermediate link in the causal chain from *P* to P^*" (1993c, 207). However, this argument makes it possible to refute scenario (3) only within the framework of the traditional theory — which we reject — according to which all determination is causal. Kim's observation that *P*

56 For a defence of this analysis of causation, see Kistler (1999b, 2006a, 2006d).

57 I limit myself here to the consideration of classical chaos. Taking into account quantum mechanics, which predicts the existence of absolute limits to the precision of measurements, according to so-called uncertainty relations, raises problems beyond the scope of this book.

cannot be a cause of *M* is correct. However, this does not refute schema (3) as such, only one of its possible interpretations. It refutes

> (3a) *P* causes *P** by causing *M*.

But it leaves open the possibility that

> (3b) *P* causes *P** by determining in a non-causal manner property *M*, which causes *P** by virtue of a psychophysical causal law,

correctly represents the situation.[58] (3b) is equivalent to the first interpretation of hypothesis (1). Kim neglects this possibility because he follows the empiricist tradition according to which there are only two ways that properties *M* and *P* can be nomologically correlated: either by a causal law or by identity. After objecting to Searle's conception of the relationship of *P* to *M* as a causal relationship, Kim concludes that the only alternative that avoids the problem of overdetermination is the hypothesis that they are identical (1998, 48).

The second argument against scenario (3) is as follows: by virtue of an "inference to the simplest explanation"[59] that Kim has already followed in arguing against hypothesis (1), the assumption that scenario (3) describes the situation correctly must give way to the simpler explanation according to which *P* causes *P** without any intervention by *M*. However, this explanation, whose greater simplicity is indeed undeniable, works only if "there is an appropriate law connecting *P*-instances with *P**-instances" (Kim 1993c, 207). The existence of such a law is not guaranteed a priori. Kim argues that the principle of closure of the physical domain guarantees the existence of a purely physical causal relationship between *P* and *P**. According to this principle, every physical event has a complete physical cause at every instant preceding it. Applied to *P**, this principle guarantees the existence of a complete physical cause at the instant of *M*, namely *P*.

Kim's third argument refutes scenario (3) by arguing that it is inadequate by virtue of "the problem of causal-explanatory exclusion" (1993c, 207).[60] If

58 This possibility is also considered by Marras (2000); Crisp and Warfield (2001); Jacob (2002).

59 Kim himself does not use this expression.

60 Kim (1989b, 1990). According to Schröder, "the place for downwards causation is the relatedness of the parts" (1998, 446) of a system. Like Kim, Schröder defends the thesis that the

there is a causal determination relationship between P and P^*, then the causal explanation that it makes possible "excludes" the adequacy of scenario (3). By virtue of causal-explanatory exclusion, two complete causal explanations cannot coexist in parallel, at least not systematically. Several objections can be raised against this argument. First, the physical parts of the brain that possess P and M have physical properties; however, the hypothesis of the existence of a purely physical law connecting P and P^* begs the question against my hypothesis according to which the physical properties P have causal effects only through the intermediary of the global properties that emerge during their interaction, which can be mental properties M. Second, even if a law $P \rightarrow P^*$ exists alongside the law $M \rightarrow P^*$, the conclusion that M lacks causal efficacy still depends on the controversial thesis that there is no systematic overdetermination of events that have a mental cause. Third, the appeal to simplicity does not in itself settle the matter one way or the other. Even if an explanation of P^* in terms of P were possible, it seems to be plausible that it would be far more complex than the explanation that appeals to M and the psychological law $M \rightarrow M^*$. Rather than justifying Kim's conclusion, the appeal to the simplicity of explanation provides a strong argument for the realism of the mental properties (see Rey 1997).

It is interesting to compare my critique of Kim's argument with a suggestion by E.J. Lowe for making the principle of physical closure compatible with the causal influence of the mind on the physical world. Lowe shows that the causal inefficacy of the mind can be concluded only from the assumption that all physical events, *at every instant preceding them*, have a sufficient physical cause.[61] However, a weaker principle according to which "every physical state has a fully sufficient physical cause" (Lowe 2000b, 30) is compatible with a

overall capacity of the system cannot be the cause of the system's evolution, for the reason that this role is played by the relational properties of the parts (mentioned in the premises of the synchronic explanation of higher-level properties). While acknowledging that interactions among the parts can determine the evolution of the system, Schröder excludes the possibility that this determination necessarily involves the synchronic determination of the overall properties of the system. "It is not the influence of a macro-property itself, but of that which gives rise to the macro-property, *viz.*, the new relatedness of the parts" (1998, 447). Schröder's thesis that no emergent macroproperty of a system is causally efficacious, only the "relatedness" of its parts, has the consequence that all macroproperties are epiphenomenal, whereas Kim (1997b, 1998) is concerned to limit this verdict to mental and other "functional" properties. See below.

61 Lowe (2000b, 29–32). See also Papineau (1993, Chapter 1); Lowe (1996, 2000a). In fact, Lowe uses a weaker principle compatible with the existence of events that have no cause. In his

dualist interactionist conception[62]: even if P^* has a sufficient physical cause at some moment that precedes it, say *P*, it is possible that the causal chain from *P* to P^* passes through intermediate mental steps. In the scenario considered by Lowe, *P* (at t_1) causes *M* (at t_2), which causes P^* (at t_3). He explicitly states that this scenario fits in well with the emergentist thesis (in the sense of diachronic emergence, which has it that the evolution of the human mind began with a purely physical state (*P*) and begins, at some moment, to give rise to mental states *M* that can then influence physical states (P^*). The problem with this scenario — which Lowe does not consider — is that it seems to lead back to the difficulty highlighted by Kim. It seems to be difficult to admit that the physical processes between *P* and P^* are, so to speak, interrupted *at the physical level*: for someone following the evolution of the brain on a purely physical or neurophysiological level, the influence of *M* on P^* would appear to be a mysterious intrusion.

Conversely, my way of reconciling the causal efficacy of mental properties with the causal closure of physics does not encounter this difficulty. Unlike Lowe's scenario, my suggestion is that an event with property *P* causes an event with P^* without making the dualist hypothesis that some events in the intermediate causal chain are purely mental.

The contribution of *M* might be indispensable from the point of view of causal responsibility. The event that has *M* also has physical properties *P*. However, there might be no law at the physical level according to which *P* directly determines P^*. Rather, *P* makes *M* emerge as a global property of the person, by virtue of the interactions between the parts of the system (where *P* is a micro-based property determined logically by the properties of the parts of the system and their relations). Causal responsibility for P^* is shared between *P* and *M*; there is a psychological *ceteris paribus* law that imposes on the system at the moment of the event P^*/M^* the constraint of possessing the property M^*. It is the state *P* that determines which one of the possible physical situations that give rise to M^* is realized. Similar scenarios are scientifically plausible in many contexts that do not involve mental causation.

---

formulation, "at every time at which a physical state has a cause, it has a fully sufficient physical cause" (2000b, 27).

62 Lowe (1993, 1996) uses mental causation to argue in favour of interactionist dualism. I submit his argument to a critical analysis in Kistler (2005b).

As I have pointed out above, the state of a physical system can be known only with finite precision. Even with perfect knowledge of the laws governing the dynamics of a chaotic system, the best available knowledge of the state at t is not sufficient for deducing the state of the system for times later than t^* with arbitrary precision. If we ask for the causally responsible state for the state P^* of the air molecules above Paris at t^*, specified with some margin of error, then there is always some time t earlier than t^*, such that there is no description of the state of the air molecules at t precise enough and knowable in principle that would be sufficient to predict P^* with a precision that lies within that margin of error. In this sense, if the state P^* at t^* is given with some finite precision, then there is no knowable-in-principle state at t causally responsible for P^*.

From his refutation of the first three scenarios, Kim concludes that (4) describes the situation correctly. This scenario presupposes — as does scenario (2), the argument that Kim offers against scenario (1), and the second and third arguments that he offers against scenario (3) — that there is a law and therefore a direct causal determination relationship between P and P^*. Kim justifies this presupposition with a principle that he considers to be part of "the natural picture for the layered physicalist world" (1993c, 208). According to this principle, "all causal relations are implemented at the physical level, and the causal relations we impute to higher-level processes are derivative from and grounded in the fundamental nomic processes at the physical level" (208). According to an equivalent version of this principle, which expresses it in terms of the causal powers of properties, "if M is instantiated on a given occasion by being realized by P, then the causal powers of *this instance of M* are identical with (perhaps, a subset of) the causal powers of P" (208). However, this "causal inheritance principle" (Kim 1998, 54) applies only to mental properties if we accept the eliminativist and apparently paradoxical assumption (developed in Kim 1998) that there are no mental properties: in fact, according to Kim, there are only physical properties but two kinds of concepts. Physical concepts correspond directly to properties (physical properties, but this specification is redundant since there are only physical properties), whereas mental concepts are second-order concepts that designate physical properties by means of quantification.

I will analyze this conception in a moment. My aim is to defend the plausibility of scenario (3b) — equivalent to the second interpretation of scenario (1) — by showing, contrary to Kim, that the relationship between first-order properties and second-order concepts of these properties is not an adequate model of the relationship between physical properties and mental properties. The causal inheritance principle is plausible only in the context of the assumption that the relationship between the physical and the mental is equivalent to the relationship between first- and second-order predicates. Without the causal inheritance principle, Kim has no argument for his claim that, at each instant preceding the event *e* that is P^*, there necessarily exists a physical fact, that *c* is *P*, such that there exists a physical law between *P* and P^* and therefore a relationship of causal responsibility that makes it possible to explain and predict P^* at the physical level on the basis of *P*.

4. Mental Properties or Physical Properties Conceived with Mental Concepts?

According to Kim, there are, strictly speaking, no mental *properties*. What we mistake for mental properties are mental *concepts* that apply to physical properties. If this thesis is correct, then the issue of the downward causal influence of mental properties no longer arises: there are no such mental properties with causal powers.

Kim's suggestion would solve the general problem of downward causation only if all higher-level properties could be construed as functional properties. Kim himself (1997b, 1998, 84–85) denies that mental properties belong in this respect to the same category as other macroscopic properties. On the contrary, he insists on the importance of the difference between two distinctions: on the one hand, the distinction between the macroscopic properties of a complex object and the microscopic properties of its parts; on the other hand, the distinction between the first-order properties — microscopic or macroscopic — of a given object and the functional properties of that object. What appear to be functional properties are really just second-order concepts (expressed by second-order predicates) that quantify over first-order properties.

Like any first-order property, macroproperties of macroscopic objects have causal powers, which can differ from the causal powers of the properties of the objects' parts. There is no principle equivalent to the causal inheritance principle that would deprive macroproperties in general of any proper causal efficacy, over and above that of the microproperties from which they emerge. The emergentist thesis, according to which mental properties are macroproperties, makes downward causation, as in scenario (3b), conceivable. Each instance of *P* determines an instance of *M*, thanks to a non-causal law of composition. If *M* is causally responsible for *P** thanks to a causal law, possibly in conjunction with property *P* underlying *M*, then we have a case of downward causation.

Conceiving of mental properties as macroproperties of macroscopic objects, in an analogy with physical, chemical, or biological macroproperties that emerge by virtue of non-linear laws of interaction that apply to the properties of their parts, makes their causal efficacy conceivable. Let us take the example of a laser, which has emergent causal powers. First and foremost, it has the power to emit coherent light (i.e., light whose components are all in phase). A laser causes a beam of extremely monochromatic light; for the first historical laser, the ruby[63] laser built in 1960, this wavelength is 6,943 Å (or 694.3 nm). To say that this light is "extremely monochromatic" means that the deviations from the average wavelength of the light emitted are very small compared with this wavelength; in this case, these deviations are of the order of 0.1 Å. This phenomenon is extraordinary in that all other natural bodies emit radiation distributed over a broad spectrum of different wavelengths, which can be modelled by Planck's law of "black body" radiation. A very specific configuration of atoms in the ruby crystal gives this crystal — as a complex object — the causal power to produce a characteristic beam of coherent, monochromatic light. There is a law of composition, according to which a certain atomic configuration in a ruby crystal determines the structure of the energy levels of the electrons of the Cr^{3+} ions of the ruby crystal that, in specific circumstances,[64] is causally responsible for the laser emission. This law is deducible, in principle, from quantum mechanics in a way analogous to the deduction of the stability of the H_2^+ ion that we considered in Chapter 4.

To see the analogy between the structure of causal determination of light emission in a laser and that of causal determination by a mental property, let us call *P* the set of properties of the atoms making up the ruby crystal alongside their spatial relationships. Let us call *C* the (chemical) properties that belong to the crystal as a complex object and are responsible for the laser

63 A ruby is composed of 99.95% Al_20_3 and 0.05% Cr_2O_3.

64 Electrons need to be excited or "pumped" to certain of these energy levels using a mercury lamp that surrounds the ruby crystal.

mechanism. *C* includes in particular the structure of the energy levels of the electrons of the Cr^{3+} ions, as it exists inside the crystal.

The operation of a laser is a case of downward causation from the chemical level to the atomic level: *P* determines *C* by virtue of a law of composition, itself determined by the laws governing the interactions between the electrons and protons in the crystal. According to a physico-chemical causal law, *C* has the causal power to produce, in precise circumstances, a physical effect *P**, namely the emission of coherent, monochromatic light.

If this is indeed a case of downward causation, then it can be analyzed according to one of the scenarios described above to elucidate the causal relationships between mental properties *M* and *M** and physical properties *P* and *P**. The only difference is that chemical properties *C* take the place of mental properties *M*. The four hypotheses considered by Kim for the case of mental causation correspond to four analogous hypotheses about the respective causal contribution of the physical properties *P* and chemical properties *C* of the ruby crystal in the production of laser light *P**:

(1) *C* and *P* together constitute a sufficient cause of *P**;

(2) *C* and *P* are two distinct sufficient causes of *P**, which overdetermine *P**;

(3) *P* causes *P** via *C*;

(3a) *P* causes *P** by causing *C*;

(3b) *P* causes *P** by determining in a non-causal manner property *C* that causes *P** by virtue of a chemical-physical causal law;

(4) *P* causes *P** directly, without any causal contribution from *C*.

Before examining these hypotheses, it is important to note that the production of monochromatic light by the ruby laser is not perfectly analogous to the process of mental causation that I analyzed above. The effect of the laser — the ray of monochromatic light — is a purely physical phenomenon in the sense that the events that constitute it have no chemical properties. This makes a distinction analogous to the one that I made in my interpretation of hypothesis (1) inappropriate; the interpretation of (1), according to which *P* and C act as two factors neither of which alone is sufficient to produce P^*, but which produce P^* jointly, does not hold here. In fact, in the case of the laser, there is no chemical law $C \rightarrow C^*$ analogous to the psychological law $M \rightarrow M^*$, the consequence of which constitutes a constraint that limits the degrees of freedom of the effect P^*.

The only possible interpretation of (1) is that P determines C non-causally and that C determines P^* causally by virtue of a law $C \rightarrow P^*$. Therefore, the only sense in which P and C together constitute a sufficient condition for P^* is that P is nomologically sufficient under the circumstances for P^*, thanks to the intermediate determination of C.

Assumption (3a) is inappropriate because the determination of C by P is non-causal. In particular, there is no time lag between the instantiation of C by the crystal and that of P by all of its atomic components.

For reasons already explained, it is empirically possible that hypotheses (2) and (4) are also inappropriate. This is the case if there is no causal law $P \rightarrow P^*$ at the level of atomic physics that would determine the effect P^* directly, without passing through the chemical level C. In this case, the effect P^* is not directly determined causally at the atomic level.

The concept of deterministic chaos allows us to understand the possibility that the evolution of a purely material system, subject only to deterministic laws (this is, I suppose, the case of the human nervous system), might be such that, if its microscopic state P^* at t^* is given with some finite precision, then there is some time t earlier than t^* such that there is no knowable-in-principle microstate P of the system at t causally responsible for P^*.

We can make the hypothesis that the brain is chaotic at the level of the properties of its neuronal components. However, the configuration of the activity of neurons and their connections determines — in a non-causal way — a global property (in reality many insofar as there are specialized cognitive modules) that is mental. This property evolves according to psychological laws. Accordingly, the fact that the cognitive system possesses certain mental properties can be causally responsible for cognitive or behavioural facts. In my example, my thought that the noise in the street is disturbing my concentration (M) leads me to make the decision to close the window (M^*). More precisely, the fact that I have the property M at t is causally responsible for the fact that I have the mental property M^* at t^*. A causal law determines the evolution of the system as a function of a global property. However, if the state of the set of all neurons is given with some finite precision, then for some t^* later than *t* the lawful evolution of the neurons does not determine their state at t^* with the same precision. The state of the set of neurons at *t* does determine their state at t^*, but this state is determined — at least in part — by the overall psychological property (or properties). In a similar way, it is indeed the set of atoms that makes up the ruby crystal that determines (and indirectly causes), in favourable circumstances, the emission of laser light, but these atoms are not directly causally responsible for this effect. They act in this way only by virtue of their interaction, which produces an overall property of the crystal: the structure of the energy levels of the Cr^{3+} ions. It is this overall property that regularly and causally determines the effect of the laser light emission, according to a causal law. In both cases, there is downward causation because a global property of the entire system causally determines a subsequent state of the system, which is situated at the level of the properties of the components (as in the case of the laser) or which constrains the system at the level of the properties of its components (as in the case of mental causation).

This suggestion bears only a superficial resemblance to a conception of downward causation suggested by Popper. In his words, "the randomness of the movements of the elementary particles — often called 'molecular chaos' — provides, as it were, the opening for the higher-level structure to interfere. A random movement is accepted when it fits into the higher-level structure; otherwise it is rejected" (1977, 348). According to this suggestion, chaos would lead to the exploration of different types of evolution, only some of which conform to higher-level laws. The non-conforming courses of events would be eliminated, in an analogy with living beings resulting from harmful genetic mutations. However, this "downward causation" brought about by natural selection differs from the "downward causation" that I have sketched above. In the case of the laser — and, as I assume, in the case of mental causation — the constraint imposed on the lower-level property by the higher-level property is immediate and deterministic. The application of Popper's model of downward causation to physical causation would imply the existence of contradictory situations in nature: for a certain period of time (i.e., before they are eliminated), there would be molecular movements that contradict higher-level laws.

Another reason for rejecting the analogy with natural selection is the fact that the selective forces at work in natural selection do not operate directly in a downward fashion. An organism poorly adapted to its environment is not breaking any laws. The forces by which the environment eliminates poorly adapted organisms are physical. It is only a question of downward causality in the sense that the presence of certain types of individuals can be explained only by higher-level laws. Campbell (1974) offers the example of the jaws of soldier termites, so large that they make these termites unable to feed. Their presence can be explained only by appealing to the laws of social organization through the division of labour in social species. However, this is a long-term explanation that does not imply any downward physical causation in the short term. The selection model cannot be transferred to downward causation. Microscopic movements have no "elbow room" to try out, even for short periods, movements that violate higher-level laws.

(3b) therefore appears to be the correct analysis. The "problem of causal-explanatory explanation" does not arise because the causal determination of laser light operates only in a downward manner. Similarly, that problem does not arise in the case of mental causation. In the absence of laws that directly determine the neuronal state P^* as a function of the previous neuronal state P, the causal determination of P^* is partly downward. The state of the set of neurons and the state of their relationships (above all their synaptic relationships) P do not directly determine the next state, P^*, of the same set of neurons; they determine it only through the intermediary of the determination of a global mental property M of the whole organism.

It is true that M "inherits" in a sense its causal powers from the underlying properties P that determine its existence. However, the metaphor of inheritance is itself misleading, and Kim is wrong to interpret it literally. For him, the "causal inheritance principle" (1992a; 1993c, 208; 1998, 54) implies that the donor has a property, with its causal powers, at the physical level. This property is transmitted unchanged to the heir (the mental level). At the mental level, we find the same physical property, now referred to with mental concepts, which has conserved its physical causal powers (Kim 2002, 674). According to my model, a better analogy to the influence that the physical level exerts on the mental level is the transmission of ideas during a conference. The speaker's words do determine the listeners' ideas. However, aside from cases in which the listener memorizes the speaker's words identically, the listeners' ideas are not identical to the speaker's ideas.

Kim's assertion that "higher states are to inherit their causal powers from the underlying states that realise them" (1993c, 208) might be true if interpreted in the sense of cultural inheritance: where the heir receives something different from what is transmitted by the donor. Kim's "causal inheritance principle" seems to be plausible only in the context of a conception of mental properties according to which they are physical properties conceived of by means of second-order concepts, which contain a quantification over these physical properties. Insofar as we construe the relationship between *P* and *M* as one of nomological determination of a property *M* of the organism by different properties of its parts — in an analogy to the determination of *C* by *P* in the case of the laser — it is not true that "the causal powers of *this instance of M* are identical with (perhaps, a subset of) the causal powers of *P*" (Kim 1993c, 208). Neurophysiological properties *P*, situated at the level of neurons and their relationships (in particular, synaptic relationships), do not belong to objects at the same level as mental properties. The relationship between the parts and the whole is a matter of (non-causal) nomological determination between *different* properties, which is not the same as realization, a relationship between different predicates referring to the *same* properties.

5. Conclusion

We have the intuition that our decisions, through our actions, can change the course of events. Philosophy alone cannot determine whether this is justified or illusory. Concepts alone cannot establish that the mind has causal powers of its own. But philosophy can determine whether this is at least conceivable. Its role is to map out the conceptual terrain and indicate the questions that only science can answer. Chapters 3 and 4 elaborated a conception of emergent systemic properties that can be applied to mental properties. In this framework, representations and decisions — alongside other mental states, processes, and events — can be construed as emergent properties that, though determined by underlying physical properties, have powers of their own. The mind is not just an epiphenomenon of the brain, as seems to follow from the functionalist conception of mental properties. It is true that many cognitive concepts are functional: that is, their logical form is second-order because it contains an existential quantification. It does not follow, however, that the only efficacious properties are physical: the first-order properties over which the second-order functional concepts quantify can be emergent properties, in particular cognitive properties.

However, there are reasons to doubt that cognitive properties can be efficacious, even if we accept the idea that they are first-order properties. Physicalism — which I have accepted as the framework for my inquiry

— implies that all real states, processes, or events are ultimately determined by physical states of affairs. True, if mental properties are first-order, then their efficacy is no more doubtful than the efficacy of emergent non-mental properties, such as chemical or biological ones. But the physicalist doctrine according to which all properties ultimately are determined by the physical properties of elementary particles seems to challenge the very intelligibility of the causal efficacy of non-physical properties. However, we have seen that physicalism is compatible with a conception of complex systems in which their non-physical properties contribute to determine their evolution; therefore, such properties have causal powers of their own. Of course, philosophy can only establish that this is rationally conceivable. It is up to the empirical sciences to establish that this or that particular emergent property is real and subject to laws of nature.

The scenario that I have developed is as follows. The determination imposed on a complex system by microscopic laws governing the behaviour of its components is articulated with macroscopic laws. Complex systems have emergent properties subject to laws that constrain the system as a whole. However, these systemic laws do not determine its evolution at the level of its microscopic components. The evolution of the system at the microscopic level is the result of a double determination: systemic laws impose constraints limiting the possibilities of evolution of the components. Within the framework of these constraints, their evolution is determined by microscopic laws. For example, the psychological law discovered by Rescorla and Wagner determines the progress of learning by classical conditioning at the level of the cognitive system. Like other laws in the special sciences, psychological laws are *ceteris paribus* laws that apply only under very specific conditions. This law does not apply in exceptional circumstances; for example, brain trauma interrupts learning by causing retrograde amnesia. Under normal conditions, however, all animals capable of conditioning evolve in such a way that their behavioural dispositions obey Rescorla and Wagner's law of conditioning. However, this law does not determine the details of the modification and formation of synapses at the physiological and molecular levels. Physiological, chemical, and physical laws determine the precise sequence of microscopic changes that give rise to learning at the systemic level. Similarly, thermodynamic or hydrodynamic laws determine the evolution of liquid or gaseous bodies at the macroscopic level without determining the microscopic evolution of their components. The detailed evolution of these components is determined by physical and chemical laws within the constraints fixed by macroscopic laws.

This model shows how an emergent property can exert a causal power of its own. It can contribute to determine the evolution of the system without calling into question either the principle of causal closure of the physical domain or the principle of causal-explanatory explanation. Let us consider the latter principle first. If system *s* has an emergent property G^* at t^*, then there are generally two ways of fully explaining G^*. G^* can be the conclusion of two kinds of deductive-nomological arguments. G^* can be explained in a non-causal way, on the basis of laws of composition and microstructural properties P^* that the system possesses at t^*, and G^* can be explained causally, by dynamic laws based on the properties that the system possesses at *t*, some time earlier. In accordance with the principle of causal-explanatory exclusion, there is only one complete causal explanation of G^*. However, there is no reason to accept the stronger "principle of explanatory exclusion." Even if the causal explanation of G^* on the basis of the state of the system at *t* is complete, there can be another complete explanation of G^* that is non-causal. This is the explanation of G^* by the microscopic state P^* of the system at time t^*.

The existence of emergent properties, sources of downward causal determination, might also seem to be incompatible with the physicalist principle of the causal closure of the physical domain. According to this principle, the physical state P^* of any system at a given instant t^* is determined, with respect to any instant *t* prior to t^*, by its physical state *P* at *t*. My model is compatible with this principle. Let us assume that P^* is causally determined partly by emergent properties *G* (and macroscopic laws) that the system possesses at *t* and partly by microscopic properties *P* that it possesses at *t*. Even if *G* makes an essential contribution to the causal determination of P^* (and G^*), ultimately it is the physical properties *P* of the system at *t* that determine P^* (and G^*). *P* directly determines, in a non-causal way, the emergent properties *G*, which then contribute — in the way indicated above — to determining the state of the system at t^*. It therefore appears that physicalism, which requires all determination ultimately to be physical, is compatible with the existence of emergent properties having causal powers of their own. The state of some complex systems is only indirectly determined causally by their previous physical state: causal determination is achieved through emergent properties and systemic laws constraining the evolution of these emergent properties.

Conclusion

The main objective of my inquiry has been to assess the intuition that our minds are real, in the sense of being able to intervene causally in the course of natural events. I have constructed a conceptual framework that enables us to justify the conviction that cognitive properties are real. It is unavoidable to express myself in such a cautious way because we have seen that conceptual work is not sufficient on its own to establish the reality of cognitive properties. I will have achieved my goal if I have shown that it is conceivable that the mind is real in the same way as the human body or any other material object.

It might seem surprising to devote so much effort to justify such a modest conclusion. Indeed, nothing seems to be as certain to common sense as the reality of the mind. Similarly, in the philosophical tradition, the reality of the mind is often considered to be at least as certain as that of material objects. However, the enormous success of science in the systematic explanation of natural phenomena gives us solid reasons for considering the reality of matter to be established with certainty. More precisely, the doctrine that I have presupposed in this inquiry, and that has provided the metaphysical framework for my analysis, is "physicalism." This term is preferable to "materialism" because it better reflects the rational basis for the choice of this metaphysical framework. That the various scientific disciplines, and above all physics as the fundamental science, are concerned with material objects and their properties is something that we have discovered empirically. It could have been that the scientific theories most successful in explaining and predicting phenomena are theories of mental phenomena. Physicalism is the metaphysical doctrine according to which science as a whole reveals what exists. The unification of the different scientific disciplines, which has played an important role in our thinking about the place of the mind in nature, gives physics a foundational place: scientific research gives us strong reasons to believe that the objects of sciences other than physics are composed exclusively of physical parts. The results of sciences such as chemistry and biology give us every reason to believe that chemical substances and living beings are composed exclusively of physical parts. Vitalist doctrines, according to which living beings also contain non-material components, such as élan vital or entelechy, have been refuted.

Physicalism results from the choice of science as a guide to ontology. Of course, science cannot refute skepticism; in other words, it cannot demonstrate the reality of the objects and properties that it investigates or of the laws that it discovers. Even less so can it demonstrate that what is not the subject of any science is not real. Finally, history shows us that the sciences are not immune to changes, sometimes radical, that lead to upheavals in the entities, objects, and properties in whose existence science gives us reason to believe. Therefore, it would be unreasonable to expect science as it stands today to give us definitive information about what exists. Certain substances, such as the phlogiston of medieval alchemy and the ether of pre-relativistic physics, have been eliminated from the set of entities for whose existence science is "ontologically committed," to use Quine's (1948) expression. Others have joined the set of objects or properties in whose existence science gives us reason to believe, such as quarks and the strong interaction force of nuclear physics, the olinguito (*Bassaricyon neblina*), a mammal discovered in 2013 in South America, or iconic memory, discovered during cognitive psychology experiments in 1960.

With regard to the particular limits of scientific knowledge, physicalism is a more cautious metaphysical doctrine than traditional materialism. Instead of adopting the dogmatic position that "everything that exists is material," physicalism holds, first, that all objects that exist are either among the objects studied by physics or composed exclusively of objects studied by physics and, second, that all real properties are either among the properties studied by physics or reducible to them.

The debate about the reality of the mind revolves around the question of the reducibility of psychological laws. Property dualists and eliminativists maintain that these laws are irreducible. However, they do not agree on the conclusion to be drawn from this thesis. Dualists conclude that the irreducibility of psychological laws and the mental properties to which they relate shows that psychology is indeed an autonomous science that could never be replaced by neuroscience. Eliminativists conclude that the irreducibility of the mind demonstrates the radical falsity of psychological statements, at least as far as common-sense psychology is concerned. This falsity entails the non-existence of the properties to which its predicates seem to refer. Reductionists rely on the successes of cognitive neuroscience to argue that at least some cognitive abilities, such as long-term memory, are reducible to neuroscience and maybe even to biochemistry. Here again, however, it is possible to draw very different conclusions. Reductionists such as Schaffner (1993) and Bickle (2003) argue, against eliminativism, that the reduction of a cognitive property establishes its reality. But their conception of reduction leads them to deny that reduced properties have a reality distinct from the reducing properties: according to these authors, reduction shows that the reduced property is *identical* to the reducing properties. It is only insofar as the fixation of memory is identical to a neurophysiological (or biochemical property) that we are justified in considering it real. On the contrary, I have defended the thesis that the reduction of a property does not imply its identification with a property at the reducing level. In particular, I have shown that it is possible to reduce multi-realizable properties. Systems of different physical natures can possess a property such as temperature or the ability to learn by conditioning. Since different properties can give rise to a given emergent property, it cannot be identical to any of the reducing properties.

The whole debate on the reality of properties outside the realm of fundamental physics, and in particular on the reality of mental properties, presupposes a fundamental conviction that also motivates the adoption of physicalism. Dualism, eliminativism, and reductionism all agree in accepting the thesis that science is the ultimate judge of what is real, regardless of whether or not they support the identity of reduced and reductive properties. Thus, what justifies considering cognitive properties as real is the discovery of laws of nature that relate to these properties. Insofar as Rescorla and Wagner's (1972) law describes a regularity in the dependence of the increase in associative strength between a conditioned stimulus and a response on different parameters — such as the salience of the unconditioned stimulus or the strength of association already present — we are justified in believing that learning by conditioning really exists as a cognitive process. The same applies to the properties of such learning, such as the increase in the strength of association as a function of the number of exposures to the stimuli.

The physicalist conviction rests on the observation of the progressive unification of the sciences. This is achieved largely through reductions between theories that study phenomena at different levels. The nature of reduction is controversial, and the first two chapters of this book were devoted to its detailed analysis. The model that I have developed is the result of a synthesis of two conceptions of reduction. From Nagel's (1961) model, I have kept two theses. First, the reduction of a property involves deducing the laws that apply to it. For example, the reduction of temperature involves deducing the laws of thermodynamics that relate temperature to other macroscopic quantities, such as pressure or entropy. Second, this deduction necessarily involves laws of composition, which play the logical role of Nagel's "bridge laws" or "linking principles." Analysis of the reductions accomplished in the history of science shows that these laws of composition are not derived a priori from knowledge of the reducing level alone. The laws of composition are always constructed on the basis of prior knowledge of the two theories unified by the reduction. The laws of thermodynamics were not discovered on the basis of the laws of classical mechanics; they were first discovered on the basis of the observation of macroscopic phenomena: that is, in the context of research carried out at their own level independently of any consideration of the microscopic level. Similarly, the laws of learning by classical conditioning were first discovered by investigating the regularities observable at the level of cognitive systems — both animal and human — without regard to their microscopic neuronal and molecular constituents. It was only once the theory of macroscopic phenomena had been developed that the search began for regularities among the microscopic components of the objects of macroscopic theory — regularities that could serve eventually as premises in a deduction of macroscopic laws.

It might be objected that this is merely a matter of the contingent order of the acquisition of knowledge and does not contradict the thesis contained in the CHB reduction model (named after Churchland, Hooker, and Bickle), according to which the deduction of macroscopic laws can be achieved in principle merely by the conceptual analysis of a microscopic description. I have provided two responses to this objection. First, it seems to be convincing only insofar as ontology is not clearly distinguished from epistemology. On the ontological level, the physicalist framework of my investigation guarantees that macroscopic phenomena and laws are determined exclusively by microscopic states of affairs and the laws that apply to them. However, the difference between the CHB thesis and the thesis that the laws of composition are discovered only through knowledge of the laws at the macroscopic level is epistemic. Once this distinction has been made clearly, the applicability of the CHB model presupposes a possibility in principle that does not correspond to any historical reality: that of deducing a priori, from knowledge of the microscopic level alone, all of the laws of the macroscopic level. The thesis that this is possible in principle cannot be refuted directly; however, the burden of proof is on those who claim something to be possible that has never been achieved. Second, I proposed that there are cases of historical reductions in which the laws of composition include a part irreducible to the laws governing microscopic phenomena. The concept of an ensemble in the sense of Gibbs (1902) has no equivalent in the microscopic description of gas molecules. Without this concept — or others just as irreducible to the molecular level — it is impossible to deduce the macroscopic laws of thermodynamics. Macroscopic quantities such as temperature can be deduced only from the average of the squares of the velocities of all molecules. However, the average over time corresponds only to a real property of the system if that system is in equilibrium. Moreover, it is impossible to derive the fact that the system is in equilibrium from knowledge of the microscopic level alone.

My synthetic model of reduction also takes into account an important criticism of Nagel's (1961) model of reduction. Research on historical reductions has shown that a reduction is generally accompanied by corrections to the reduced theory. These corrections are even the main motive for the search for reductions. To account for the difference between the theory shown to be deducible from the reducing theory, and the reduced theory as it was before the reduction, Schaffner (1967) introduced the concept of positive analogy: the theory T_R* that can be deduced from the reducing theory T_B is not identical, but only structurally analogous, to the reduced theory T_R. The CHB model takes account of the difference between the theory that needs to be reduced, T_R, and the theory T_R* that can be deduced from the reducing theory T_B. But the CHB model accompanies the recognition of this difference with a thesis that I have rejected: according to the CHB model, the deduction of T_R* from T_B is a case of *intratheoretical* deduction insofar as this deduction does not require any concept or principle external to the reducing theory T_B. The analogous relationship between T_R and T_R* then becomes the only *intertheoretical* part of the reduction.

According to the synthetic model of reduction that I have offered, it is generally necessary to use laws of composition that cannot be derived a priori from the reduction theory T_B alone. Therefore, contrary to the CHB thesis, the deduction of T_R* from T_B is not intratheoretical in T_B since it presupposes knowledge of T_R and sometimes requires recourse to new principles found neither in T_B nor in T_R. My synthetic model also recognizes — following Schaffner and the CHB model — that deduction from T_B and the laws of composition generally leads to a theory $T_R{}^*$ analogous to, but not identical with, the theory T_R that is the target of reduction.

The reduction of a systemic property is based above all on the discovery of a law of composition that determines that all complex objects with a certain structure necessarily possess this property. This discovery makes it possible to integrate the property into the system of scientific knowledge, the best way of justifying its reality. However, I have found reasons to contest the thesis defended by Causey (1977) and Schaffner (1993) according to which the reduction of a property leads to its *identification* with properties, or functions of properties, of the reductive theory. A property of a macroscopic object cannot, for logical reasons, be identical to properties of its microscopic components. The only properties with which it is logically possible to identify a systemic macroscopic property of a complex object are "structural" properties (Armstrong 1978) that Kim (1998) calls "micro-based properties." A complex macroscopic object has such a micro-based property on the mere logical basis that it has a number of parts $p_1, p_2, \ldots, p_n$, each with a number of properties $P_{11}, P_{12}, \ldots, P_{21}, \ldots, P_{nm}$, and that there are spatial relationships R among these parts. We have seen that the conditions for the existence of such a micro-based property are not sufficient to guarantee it a real existence in the sense of having causal powers of its own. Indeed, insofar as the existence of nomic interactions among the parts is not required, the mereological whole made up of my left shoe and your right shoe, or the mereological whole made up of the left hemisphere of my brain and the right hemisphere of your brain, have such micro-based properties. But in the absence of the relevant interactions, they have no real properties, in the sense of causal efficacy. No cognitive property emerges from the mereological whole made up of my left cerebral hemisphere and your right cerebral hemisphere. However, when the parts of such a mereological whole interact, it is possible that real systemic properties emerge that are qualitatively different from the properties of the parts. The laws of interaction between the atomic components of the hydrogen molecule give rise to the stable structure of the molecule. The laws of interaction that govern the interaction between the neurons and neuronal networks in my brain give rise to the thoughts that I am in the process of transcribing onto paper. If the interaction gives rise to a whole with emergent properties, then we can say that the emergent property is determined in a non-causal way by a "law of composition."

My analysis has shown that the notions of reduction and "level of reality" are crucial for physicalism. These notions are indispensable for reconciling the thesis that the microphysical level determines all real states, processes, and events, with the reality of entities that are not microphysical, where "reality" means that these entities have causal powers of their own. Physical states of affairs determine other levels of reality but only by means of laws of composition that sometimes cannot be reduced to microscopic laws alone — as in the case of the thermodynamic hypothesis of equilibrium — and that generally cannot be derived a priori from knowledge of the microscopic level alone. We can understand the relationship among the different levels of reality only on the basis of knowledge acquired through observations and experiments conducted at all levels.

But the notion of levels of reality would not be fully justified if it were only a question of classifying phenomena as objects of knowledge. The thesis that levels of reality really exist, and are not just an effect of perspective generated by our fragmented approach to reality, can be justified only within the framework of a doctrine of emergence. Chemical phenomena have their own reality and their own causal powers in relation to microphysical causal powers only insofar as they are objectively different from microphysical phenomena. My analysis of the notion of emergence in Chapter 4 has enabled us to give this notion an ontological meaning that allows us to account for the qualitative difference between the phenomena that make up the different levels of reality. Interactions among objects belonging to a given level lead to the appearance or "emergence" of complex objects with properties qualitatively different from those of their components. These qualitatively new properties justify the idea that the complex objects that possess them form a level of reality that differs from that of their parts. Atoms occupy a relatively fundamental level of reality. The interactions among atoms described by quantum physics give rise to molecules and macroscopic objects, in particular solid bodies. Solid bodies possess many new properties that the atoms of which they are composed do not: solids are hard or malleable, transparent or opaque and coloured, whereas atoms cannot have any of these properties.

Living things occupy another level that seems to be clearly distinct in its characteristic properties. Living things organize and reproduce themselves, whereas their components do not. There can be controversy about the delimitation of the different levels characterized by specific properties. Here again the main way of overcoming these controversies is scientific: the existence of a level gives rise to a set of specific phenomena that are the subject of a specific science. The very existence of chemistry attests to the fact that there is a set of specifically chemical phenomena as well as a set of regularities that chemistry takes into account. Similarly, the existence of biology gives us reason to believe in the existence of a level specific to living beings and their properties. Finally, the existence of psychology gives us reason to believe in the existence of a level specific to cognitive systems and their properties. Controversies can arise from the difficulty of judging the status of different subdisciplines and their specific objects and properties: we can debate whether plants, fluids, and atmospheres of planets constitute separate levels, because they are the subject of botany, a subdiscipline of biology, and hydrodynamics and meteorology, subdisciplines of physics. Of course, it is also essential to accompany our judgment of the existence of a level with the prudent precaution that the history of science teaches us: subdisciplines and even main scientific disciplines appear and disappear. Thus, we have reason to believe in the existence of a psychological level of reality only since the birth of scientific psychology in the nineteenth century. The analysis of the concept of reduction also shows that the reduction of the properties and theories of one level to the properties and theories of lower levels often leads to the appearance of a new discipline that appears to be hybrid from the point of view of the disciplines existing before the reduction. The reduction of an elementary part of chemistry to physics gave rise to physical chemistry; the reduction of certain cognitive abilities to neuroscience gave rise to cognitive neuroscience. It is only by looking back in the long run that we can hope to make a well-founded judgment of the nature of the levels that correspond to these sciences.

Fortunately, it is not necessary for us to enter into these controversies and answer these difficult questions. I have achieved my goal if I have succeeded in making cognition and cognitive properties appear as occupants of a distinct level of reality, in the same way as chemical and biological objects and properties. It is from this general perspective that I have answered the question of whether cognition and its properties are "nothing other" than physical objects and properties and in what sense. Are cognitive phenomena mere physical phenomena conceived differently, with a different conceptual apparatus, as is claimed by the a priori implication thesis examined in

Chapter 2 and defended for mental phenomena other than *qualia*, among others, by Chalmers (1996), Jackson (1998), and Kim (1998)? Are there mental phenomena that, conversely, are irreducible (both conceptually and empirically) to physical states of affairs and laws, as these authors maintain that *qualia* are? I have shown that negative answers to these two questions are at least consistent. It is conceivable that cognitive properties emerge from neurophysiological properties, so that they constitute their own level of reality with their own laws and causal powers, in the same way as chemical and biological properties. It is up to psychology, neuroscience, and especially the new science "between levels" — cognitive neuroscience — to show whether this conception really corresponds to the relationship between our minds and our brains.

We might consider construing levels of reality in terms of causal interactions: we might consider defining a level as the set of objects with which a given object, or a given kind of object, is capable of causal interaction. In a similar way, we might consider defining the characteristic properties of a level as the set of properties that enter into relationships of causal responsibility. Atoms interact primarily with atoms; macroscopic objects, such as our bodies, interact primarily with other macroscopic objects. However, this criterion does not lead to a clear delimitation of distinct levels because there are causal relationships between entities that intuitively belong to different levels. When a subject perceives, in a psychophysical experiment, an isolated photon absorbed by her retina, an elementary object, the photon, causes a cognitive effect situated at the level of the person. It therefore seems to be more sensible to ground the concept of level of reality on the existence of a set of properties that is the subject of a specific science and integrated into a set of laws of nature that is the object of the theories of that science.

To justify the intuition that cognitive properties are properties of persons (or animals) qualitatively distinct from the properties of their parts, we need to show that they are emergent. Therefore, it is important to find a criterion of emergence in the ontological sense. We have seen that physicalism imposes a certain number of necessary conditions for emergence. Emergent properties belong to objects composed entirely of physical parts, and they are determined exclusively by the physical properties of these parts and their interactions. The emergent properties of an object are systemic in the sense of not belonging to the parts of the object. However, these are only the necessary conditions for emergence that are also satisfied by properties not intuitively emergent: the property of a stone of weighing 5 kg does not seem to be emergent, even though none of its parts has it and even though it is determined exclusively by the properties of the parts of the stone.

I have suggested that there are mathematical criteria that can yield a sufficient condition for emergence at least for certain types of systems. For emergent physical properties, the topology of a system's trajectory in a phase space can provide such a criterion: when a purely quantitative change in the properties of a system's components can change the topological structure of the system's trajectory, that trajectory is emergent. The scientific discovery of this topological structure can justify the judgment that the system has an emergent property. I have shown that the topological difference between the psychological space of representation and the physical space of the represented stimuli allows us to apply this topological criterion to certain emergent mental properties. It remains to be determined whether it is possible to generalize the application of this criterion to other cognitive properties. In the meantime, the existence of nomic regularities at the cognitive level, which intuitively seem to be qualitatively different from the regularities that characterize neuronal processes, gives us reason to believe that there is indeed a distinct cognitive level of reality.

My thesis that there are levels of reality with their own causal powers faces two objections, which I considered in Chapters 3 and 5. One influential view is that mental properties are dispositions. To have learned to associate a conditional stimulus (CS) with the unconditional stimulus (US) that triggers the response R is to have the disposition to react with R to the perception of the CS. The idea that mental states correspond to functional roles is the common heritage of analytical behaviourism and functionalism. The essence of a cognitive state consists of what causes it and what it causes independently of the intrinsic structure of the cognitive system. Ever since the polemic against the occult powers of medieval philosophy, likened to the "dormitive virtues" of opium, dispositions have been taken to be properties of dubious reality. It seems to be gratuitous to postulate the existence of a dormitive virtue in opium when it comes to identifying the property that causes the smoker to fall asleep. Instead of identifying, in a scientific manner, a real and intrinsic property of opium causally responsible for sleep, the postulate of a disposition to induce sleep seems to create only the illusion of knowledge. In Chapter 3, I considered a number of traditional objections to the reality of dispositions, which are also objections to their causal efficacy. It became clear that these objections do not refute my conception of causally efficacious properties, according to which they can be construed both dispositionally and categorically. Even if many mental properties are indeed conceived according to their functional roles, there is nothing to prevent the occupants of these roles from being macroscopic mental properties.

According to the functionalist theory of dispositions, the occupants of the roles that characterize dispositions are always *microscopic* categorical properties. In the same vein, the functional model of reduction developed by Kim (1998) for mental properties proposes that the causally efficacious properties that occupy mental roles are microstructural properties. According to this model, what makes the animal conditioned to the CS react by R are neuronal and biochemical properties or "micro-based" properties, which correspond to logical constructions from microscopic properties. However, if causal efficacy lies exclusively at the microphysical level, then it follows that there are no macroscopic properties and in particular no cognitive properties. The cognitive level is merely a *conceptual* level to which corresponds no level of real properties. The mental thus appears as epiphenomenal.

We have seen that it is possible to avoid this conclusion on the condition that a clear distinction is made between two meanings of "reduction" and between two meanings of "realization." Showing that a categorical macroscopic property occupies a given functional role, or "realizes" the role, constitutes a first step that I have called "role-occupant reduction." It is only at a second step that a macroscopic property is reduced, in the sense of microreduction, to microscopic properties or mechanisms. In one sense, the occupant realizes the role. Hemoglobin, for example, performs the role of oxygen carrier in mammalian blood. But we can also say, in another sense of "realize," that the microproperties that determine a macroproperty in the non-causal sense realize it. The hemoglobin macromolecule is realized, in this second sense, by a certain chain of amino acids that are its microscopic components. Both forms of realization are compatible with multi-realizability. The function of transporting oxygen is multi-realized, in the first sense of realization, by hemoglobin, hemerythrin, and hemocyanin. Hemoglobin is a macromolecule multi-realized, in the second sense of realization, by different chains of amino acids.

It is conceivable that mental properties are emergent properties of cognitive systems that fulfill the roles defined by cognitive concepts. They can be causally efficacious even though they are determined by microscopic, neuronal, and biochemical properties and even though they can therefore be micro-reduced. In Chapter 5, I considered another important objection to the reality of mental properties: even if they are emergent and categorically conceivable, the principle of the causal closure of the physical domain and the principle of causal-explanatory exclusion seem to be incompatible with the idea that emergent properties exert their own causal influence on the course of physical events. According to the principle of causal closure, any physical event, at any previous moment, has a complete and exclusively physical cause. This principle is based on the observation that physics never discovers intrusions from non-physical causes. Therefore, at the moment when I make a decision, the physical consequences of that decision have an exclusively physical cause. If there is a complete physical cause at the moment of the decision, then what causal contribution could the decision itself make to a given physical consequence of my action caused by my decision? The principle of causal-explanatory exclusion states that, if causal "overdetermination" exists, then it is not systematic. Only in exceptional situations do two causal chains converge on the same event. This excludes the possibility that the physical causes of the consequences of our actions are systematically accompanied by parallel mental causes. From this reasoning, Kim (1998) draws the conclusion that there are no mental causes. Mental concepts are second-order concepts that quantify over first-order properties, which hold the monopoly of causal efficacy and, in his view, are always physical.

I have proposed the following way of avoiding the conclusion that there are no efficacious mental properties. I accept the principle of causal-explanatory exclusion. However, we have seen that physicalism does not oblige us to accept the principle of the causal closure of the physical domain. What is justified is a weaker principle according to which, at every instant prior to a given physical event, there exists a set of physical states of affairs that *determines* the event in question. Unlike the principle of causal closure, this determination can include a non-causal stage. Let us say that I decide to close the window, and this decision leads to the event of closing the window. There is a set of neural events underlying the decision that determines it in a non-causal way. It is conceivable that the decision contributes causally to the event of closing the window, along the lines of the following scenario. There is no microscopic law that determines at the neuronal level alone the detailed evolution of the complex system that is the person and her brain, in the sense that the evolution could be predicted or explained on the basis of the knowledge of the neuronal state with finite precision. However, the mental property of making this decision imposes a constraint on the system to evolve in the direction of an action that leads to closing the window. It is only a constraint because, first, many other mental properties exert influences on the evolution of the system, so that the regularities generated by each of these cognitive constraints are never strict, and because, second, the mental property alone does not determine the evolution of the system in microscopic detail; it determines its evolution only at the cognitive level. Each one of the stages in the preparation and execution of the action is compatible with many underlying physiological states. It is the preceding physiological state that determines which one of the physiological states compatible with a given cognitive state is actually realized. In this way, my closing the window, at the moment of my decision, has a complex cause: the decision as a mental property is indispensable to the causal determination of the action, and the underlying neurological state determines the evolution of the system among the possibilities compatible with the evolution imposed by psychological regularity. The principle of causal-explanatory closure is respected because there is only one complete explanation for the closing of the window. This explanation is partly mental and partly physiological. However, insofar as the determination of the events caused by our minds includes a non-physical aspect, the principle of causal closure is not respected. But physicalism does not require such a strong principle. My scenario is compatible with a principle of physical determination of all physical events: each physical event, at each instant preceding it, has a set of physical states of affairs that determines it entirely.

I end this conclusion with the remark with which I began it. It is up to the various relevant sciences, and above all neuroscience and psychology, to establish whether certain processes by which our minds seem to intervene in the physical world really correspond to the scenario outlined here. The justification of the reality and causal efficacy of mental properties belongs, at least in part, to science. The properly philosophical objective that I set for myself has been achieved if I have succeeded in showing that it is at least conceivable.

References

Achinstein, Peter. 1974. "The Identity of Properties." *American Philosophical Quarterly* 11, no. 4: 257–75.

Alexander, Samuel. 1920. *Space, Time and Deity.* 2 vols. London: Macmillan.

Alston, William P. 1971. "Dispositions, Occurrences and Ontology." In *Dispositions*, edited by R. Tuomela, 359–88. Dordrecht: Reidel, 1978.

Anderson, P.W. 1972. "More Is Different: Broken Symmetry and the Nature of the Hierarchical Structure of Science." *Science* 177, no. 4047: 393–96.

Antony, Louise. 1991. "The Causal Relevance of the Mental: More on the Mattering of Minds." *Mind and Language* 6: 295–327.

Antony, Louise M., and Joseph Levine. 1997. "Reduction with Autonomy." In *Philosophical Perspectives 11: Mind, Causation, and World*, edited by James E. Tomberlin, 83–105. Cambridge, MA: Blackwell.

Armstrong, David M. 1968. *A Materialist Theory of Mind.* London: Blackwell.

———. 1973. "Beliefs as States." In *Dispositions*, edited by Raimo Tuomela, 411–25. Dordrecht: Reidel, 1978.

———. 1978. *Universals and Scientific Realism.* 2 vols. Cambridge, UK: Cambridge University Press.

———. 1983. *What Is a Law of Nature?* Cambridge, UK: Cambridge University Press.

———. 1996. "Place's and Armstrong's Views Compared and Contrasted." In *Dispositions: A Debate*, by David M. Armstrong, C.B. Martin, and U.T. Place, edited by Tim Crane, 33–48. London: Routledge.

———. 1997. *A World of States of Affairs.* Cambridge, UK: Cambridge University Press.

———. 1999. "The Causal Theory of Properties: Properties According to Shoemaker, Ellis and Others." *Philosophical Topics* 26: 25–37; also in *Metaphysica* 1 (2000): 5–20.

Ashcroft, N.W., and N.D. Mermin. 1976. *Solid State Physics.* Philadelphia: Saunders College.

Baker, Lynne Rudder. 1993. "Metaphysics and Mental Causation." In *Mental Causation*, edited by John Heil and Alfred Mele, 75–96. Oxford: Clarendon Press.

———. 1998. "What We Do: A Nonreductive Account of Mental Causation." In *Human Action, Deliberation, and Causation*, edited by Jan Bransen and Stefaan E. Cuypers, 249–70. Dordrecht: Kluwer.

Balzer, Wolfgang, C.U. Moulines, and J. Sneed. 1987. *An Architectonic for Science*. Dordrecht: Reidel.

Barlow, Horace. 1972. "Single Units and Sensation: A Neuron Doctrine for Perceptual Physiology." *Perception* 1: 371–94.

Batterman, Robert W. 1995. "Theories between Theories." *Synthese* 103: 171–201.

———. 2002. *The Devil in the Details: Asymptotic Reasoning in Explanation, Reduction, and Emergence*. Oxford: Oxford University Press.

Bechtel, William. 2009. "Looking down, around, and Up: Mechanistic Explanation in Psychology." *Philosophical Psychology* 22: 543–64.

Bechtel, William, and Robert Richardson. 1992. "Emergent Phenomena and Complex Systems." In *Emergence or Reduction? Essays on the Prospects of Nonreductive Physicalism*, edited by Ansgar Beckermann, Hans Flohr, and Jaegwon Kim, 257–88. Berlin: de Gruyter.

Beckermann, Ansgar. 1992. "Introduction — Reductive and Nonreductive Physicalism." In *Emergence or Reduction? Essays on the Prospects of Nonreductive Physicalism*, edited by Ansgar Beckermann, Hans Flohr, and Jaegwon Kim, 1–21. Berlin: de Gruyter.

Bedau, Mark A. 1997. "Weak Emergence." In *Philosophical Perspectives 11: Mind, Causation, and World*, edited by James E. Tomberlin, 375–99. Cambridge, MA: Blackwell.

Bennett, Jonathan. 1988. *Events and Their Names*. Indianapolis: Hackett.

Bennett, Karen. 2003. "Why the Exclusion Problem Seems Intractable, and How, Just Maybe, to Tract It." *Noûs* 37: 471–97.

Bernal Velasquez, Reinaldo J. 2012. *E-Physicalism: A Physicalist Theory of Phenomenal Consciousness*. Frankfurt: Ontos Verlag.

Besson, Jean-Marie. 1992. La Douleur. Paris: Odile Jacob.

Bickle, John. 1992. "Mental Anomaly and Mind-Brain Reductionism." *Philosophy of Science* 59: 217–30.

———. 1998. *Psychoneural Reduction*. Cambridge, MA: MIT Press.

———. 2003. *Philosophy and Neuroscience: A Ruthlessly Reductive Account*. Dordrecht: Kluwer.

Bird, Alexander. 1998. "Dispositions and Antidotes." *Philosophical Quarterly* 48: 227–34.

———. 2007. *Nature's Metaphysics: Laws and Properties*. Oxford: Clarendon Press.

Blackburn, Simon. 1990. "Filling in Space." *Analysis* 50: 62–65.

———. 1993. "Losing Your Mind: Physics, Identity, and Folk Burglar Prevention." In *Essays in Quasi-Realism*, 229–54. Oxford: Oxford University Press.

Block, Ned. 1990. "Can the Mind Change the World?" In *Meaning and Method: Essays in Honour of Hilary Putnam*, edited by G. Boolos, 137–70. Cambridge, UK: Cambridge University Press.

———. 1997. "Anti-Reductionism Slaps Back." In *Philosophical Perspectives 11: Mind, Causation, and World*, edited by James E. Tomberlin, 107–32. Cambridge, MA: Blackwell.

Block, Ned, and Robert Stalnaker. 1999. "Conceptual Analysis, Dualism, and the Explanatory Gap." *Philosophical Review* 108: 1–46.

Braddon-Mitchell, David, and Frank Jackson. 1996. *The Philosophy of Mind and Cognition.* Oxford: Blackwell.

Broad, C.D. 1925. *The Mind and Its Place in Nature.* London: Harcourt, Brace. (Reprinted, London: Routledge, 2000.)

Bunge, Mario. 1977. "Emergence and the Mind." *Neuroscience* 2: 501–09.

———. 2003. *Emergence and Convergence: Qualitative Novelty and the Unity of Knowledge.* Toronto: University of Toronto Press.

Byrne, Alex. 1999. "Cosmic Hermeneutics." *Philosophical Perspectives 13: Epistemology*, edited by James E. Tomberlin, 347–83. Cambridge, MA: Blackwell.

Campbell, Donald T. 1974. "Downward Causation." In *Studies in the Philosophy of Biology: Reduction and Related Problems*, edited by F.J. Ayala and T. Dobhzansky, 179–86. Berkeley: University of California Press.

Campbell, H.F. 1860. "Caffeine as an Antidote to the Poisonous Narcotism of Opium." *Boston Medical and Surgical Journal* 63: 101–04.

Campbell, Keith. 1990. *Abstract Particulars.* Oxford: Blackwell.

Carnap, Rudolf. 1936–37. "Testability and Meaning." *Philosophy of Science* 3 (1936): 420–71; "Testability and Meaning II." *Philosophy of Science* 4 (1937): 1–40.

———. 1956. *Meaning and Necessity.* 2nd ed. Chicago: University of Chicago Press.

Carnot, Sadi. 1824. *Réflexions sur la puissance motrice du feu et sur les machines propres à développer cette puissance.* Paris: Bachelier.

Caro, Paul. 1995. *De l'eau: Questions de science.* Paris: Hachette.

Carroll, John. 1994. *Laws of Nature.* Cambridge, UK: Cambridge University Press.

Cartwright, Nancy. 1983. *How the Laws of Physics Lie.* Oxford: Clarendon Press.

———. 1989. *Nature's Capacities and Their Measurement.* Oxford: Oxford University Press.

———. 1999. *The Dappled World. A Study of the Boundaries of Science.* Cambridge, UK: Cambridge University Press.

Causey, Robert L. 1977. *Unity of Science.* Dordrecht: Reidel.

Chalmers, David. 1996. *The Conscious Mind.* New York: Oxford University Press.

———. 2004. "The Foundations of Two-Dimensional Semantics." In *Two-Dimensional Semantics: Foundations and Applications*, edited by J. Garcia-Carpintero and J. Macia, 55–140. Oxford: Oxford University Press. https://.

Chalmers, David, and Frank Jackson. 2001. "Conceptual Analysis and Reductive Explanation." *Philosophical Review* 110: 315–60.

Charles, David. 1992. "Supervenience, Composition, and Physicalism." In *Reduction, Explanation, and Realism*, edited by David Charles and Kathleen Lennon, 265–96. Oxford: Oxford University Press.

Churchland, Patricia S. 1986. *Neurophilosophy: Toward a Unified Science of the Mind-Brain.* Cambridge, MA: MIT Press.

Churchland, Patricia S., and T.J. Sejnowski. 1992. *The Computational Brain.* Cambridge, MA: MIT Press.

Churchland, Paul M. 1979. *Scientific Realism and the Plasticity of Mind.* Cambridge, UK: Cambridge University Press.

———. 1985. "Reduction, Qualia, and the Direct Introspection of Brain States." *Journal of Philosophy* 82: 8–28.

Churchland, Paul M., and Patricia S. Churchland. 1994. "Intertheoretic Reduction: A Neuroscientist's Field Guide." In *The Mind-Body Problem*, edited by Richard Warner and Tadeusz Szubka, 41–54. Oxford: Blackwell.

Clark, Andy. 2008. *Supersizing the Mind: Embodiment, Action, and Cognitive Extension.* Oxford: Oxford University Press.

Clark, Andy, and David Chalmers. 1998. "The Extended Mind." *Analysis* 58: 10–23.

Clark, Austen. 1993. *Sensory Qualities.* Oxford: Clarendon Press.

Cohen-Tannoudji, Claude, Bernard Diu, and Franck Laloë. 1977. *Mécanique quantique.* 2nd ed. 2 vols. Paris: Hermann.

Conway, Bevil R., Soumya Chatterjee, Greg D. Field, Gregory D. Horwitz, Elizabeth N. Johnson, Kowa Koida, and Katherine Mancuso. 2010. "Advances in Color Science: From Retina to Behavior." *Journal of Neuroscience* 30: 14955–63.

Crane, Tim. 1995. "The Mental Causation Debate." *Proceedings of the Aristotelian Society*, supplementary volume, 69: 211–36.

Crane, Tim, and D.H. Mellor. 1990. "There Is No Question of Physicalism." *Mind* 99: 185–206.

Craver, Carl F. 2007. *Explaining the Brain.* New York: Oxford University Press.

Craver, Carl F., and William Bechtel. 2007. "Top-Down Causation without Top-Down Causes." *Biology and Philosophy* 22: 547–63.

Craver, Carl F., and Lindley Darden. 2013. *In Search of Mechanisms: Discoveries across the Life Sciences.* Chicago: University of Chicago Press.

Crisp, Thomas M., and Ted A. Warfield. 2001. "Kim's Master Argument." *Noûs* 35: 304–16.

Cummins, Robert. 1983. *The Nature of Psychological Explanation.* Cambridge, MA: MIT Press.

Darden, Lindley, and Nancy Maull. 1977. "Interfield Theories." *Philosophy of Science* 44: 43–64.

Davidson, Donald. 1963. "Actions, Reasons, and Causes." In *Essays on Actions and Events*, by Donald Davidson, 3–19. Oxford: Clarendon Press, 1980.

———. 1970. "Mental Events." In *Essays on Actions and Events*, by Donald Davidson, 207–27. Oxford: Clarendon Press, 1980.

———. 1980. *Essays on Actions and Events.* Oxford: Clarendon Press.

———. 1993. "Thinking Causes." In *Mental Causation*, edited by J. Heil and A. Mele, 3–19. Oxford: Clarendon Press.

———. 1995. "Laws and Cause." *Dialectica* 49" 263–79.

Dretske, Fred. 1988. *Explaining Behavior.* Cambridge, MA: MIT Press.

Drude, Paul. 1900. "Zur Elektronentheorie der Metalle." *Annalen der Physik* 306, no. 3: 566–613.

Duhem, Pierre. 1906. *La théorie physique.* Reprinted, Paris: Vrin, 1981.

Duncan, C.P. 1949. "The Retroactive Effect of Electroshock on Learning." *Journal of Comparative and Physiological Psychology* 42: 32–44.

Earman, John, and John Roberts. 1999. "*Ceteris Paribus*, There Is no Problem of Provisos." *Synthese* 118: 439–78.

Eells, Ellery. 1991. *Probabilistic Causality.* Cambridge, UK: Cambridge University Press.

Ehring, Douglas. 1996. "Mental Causation, Determinables and Property Instances." *Noûs* 30, no. 4: 461–80.

———. 1997. *Causation and Persistence: A Theory of Causation.* New York: Oxford University Press.

Einstein, Albert. 1902. "Kinetische Theorie des Wärmegleichgewichts und des zweiten Hauptsatzes der Thermodynamik." *Annalen der Physik* 9: 417–33.

———. 1903. "Eine Theorie der Grundlagen der Thermodynamik." *Annalen der Physik* 11: 170–87.

Enc, Berent. 1983. "In Defense of the Identity Theory." *Journal of Philosophy* 80: 279–98.

Endicott, Ronald. 1998. "Collapse of the New Wave." *Journal of Philosophy* 95: 53–72.

Esfeld, Michael, and Christian Sachse. 2011. *Conservative Reductionism.* New York: Routledge.

Fales, Evan. 1990. *Causation and Universals.* London: Routledge.

Feigl, Herbert. 1958. "The 'Mental' and the 'Physical.'" In *Minnesota Studies in the Philosophy of Science*, vol. 2, edited by G. Maxwell, H. Feigl, and M. Scriven, 370–497. Minneapolis: University of Minnesota Press. (Reprinted as a monograph, *The "Mental" and the "Physical": The Essay and a Postscript.* Minneapolis: University of Minnesota Press, 1967.)

Feltz, Bernard. 1995. "Le réductionnisme en biologie: Approches historiques et épistémologique." *Revue philosophique de Louvain* 93, nos. 1–2: 9–32.

Feyerabend, Paul K. 1962. "Explanation, Reduction and Empiricism." In *Minnesota Studies in the Philosophy of Science*, vol. 3, edited by H. Feigl and G. Maxwell, 28–97. Minneapolis: University of Minnesota Press.

Fodor, Jerry A. 1974. "Special Sciences, or the Disunity of Science as a Working Hypothesis." In *Representations*, by Jerry A. Fodor, 127–45. Cambridge, MA: MIT Press, 1981.

———. 1989. "Making Mind Matter More." *Philosophical Topics* 17, no. 1: 59–79.

Funkhouser, Eric. 2006. "The Determinable-Determinate Relation." *Noûs* 40: 548–69.

Gallistel, Charles R. 1990. *The Organization of Learning.* Cambridge, MA: MIT Press.

Gardner, Howard. 1982. *Developmental Psychology: An Introduction.* 2nd ed. Boston: Little, Brown.

Gibbs, J. Willard. 1902. *Elementary Principles in Statistical Mechanics.* Reprinted, New York: Dover, 1960.

Gillett, Carl. 2016. *Reduction and Emergence in Science and Philosophy.* Cambridge, UK: Cambridge University Press.

Girill, T.R. 1976. "Evaluating Micro-Explanations." *Erkenntnis* 10: 387–405.

Glennan, Stuart S. 1996. "Mechanism and the Nature of Causation." *Erkenntnis* 44: 49–71.

———. 2010. "Mechanisms, Causes, and the Layered Model of the World." *Philosophy and Phenomenological Research* 81: 362–81.

Glymour, Clark. 1970. "On Some Patterns of Reduction." *Philosophy of Science* 37: 340–53.

Gold, Ian, and Daniel Stoljar. 1999. "A Neuron Doctrine in the Philosophy of Neuroscience." *Behavioral and Brain Sciences* 22, no. 5: 585–642.

Goodman, Nelson. 1983. *Fact, Fiction, and Forecast*. 4th ed. Cambridge, MA: Harvard University Press.

Grelling, Kurt, and Paul Oppenheim. 1937–38. "Der Gestaltbegriff im Lichte der neuen Logik." *Erkenntnis* 7: 211–25; "Supplementary Remarks on the Concept of Gestalt." *Erkenntnis* 7: 357–59.

———. 1939. "Concerning the Structure of Wholes." *Philosophy of Science* 6: 487–88.

Haas-Spohn, Ulrike. 1995. *Versteckte Indexikalität und subjektive Bedeutung*. Berlin: Akademie-Verlag.

———. 1997. "The Context Dependency of Natural Kind Terms." In *Direct Reference, Indexicality, and Propositional Attitudes*, edited by Wolfgang Künne, Albert Newen, and Martin Anduschus, 276–90. Stanford, CA: CSLI Publications.

Hall, Zach W. 1992. *An Introduction to Molecular Neurobiology*. Sunderland, MA: Sinauer Associates.

Hardcastle, Valerie Gray. 1998. "On the Matter of Minds and Mental Causation." *Philosophy and Phenomenological Research* 58: 1–25.

Hardin, C.L. 1988. *Color for Philosophers: Unweaving the Rainbow*. Indianapolis: Hackett.

Harré, Rom. 1986. *Varieties of Realism*. Oxford: Blackwell.

———. 1997. "Is There a Basic Ontology for the Physical Sciences?" *Dialectica* 51: 17–34.

Harré, Rom, and H. Madden. 1975. *Causal Powers*. Oxford: Basil Blackwell.

Hawkins, R.D., and E.R. Kandel. 1984. "Is There a Cell-Biological Alphabet for Simple Forms of Learning?" *Psychological Review* 91: 375–91.

Heil, John. 1992. *The Nature of True Minds*. Cambridge, UK: Cambridge University Press.

———. 2004. *Philosophy of Mind: A Contemporary Introduction*. 2nd ed. London: Routledge.

Hempel, Carl G. 1935. "Analyse logique de la psychologie." *Revue de synthèse* 10: 27–42.

———. 1942. "The Function of General Laws in History." *Journal of Philosophy* 39 (2): 35–48. (Reprinted in Hempel 1965a, 231–42.)

———. 1965a. *Aspects of Scientific Explanation*. New York: Free Press.

———. 1965b. "Empiricist Criteria of Cognitive Significance." In Hempel 1965a, 101–22.

———. 1966. *Philosophy of Natural Science*. Englewood Cliffs, NJ: Prentice Hall.

———. 1988. "Provisos: A Problem Concerning the Inferential Function of Scientific Theories." In *The Limitations of Deductivism*, edited by Adolf Grünbaum and Wesley Salmon, 19–36. Los Angeles: University of California Press.

———. 2002. "The Logical Analysis of Psychology." In *Philosophy of Mind: Classical and Contemporary Readings*, edited by David J. Chalmers, 14–23. New York: Oxford University Press.

Hempel, Carl G., and Paul Oppenheim. 1948. "Studies in the Logic of Explanation." *Philosophy of Science* 15: 135–75. (Reprinted in Hempel, 1965a, 245–90.)

Henle, Paul. 1942. "The Status of Emergence." *Journal of Philosophy* 39: 486–93.

Hering, Ewald. 1920. *Grundzüge der Lehre vom Lichtsinn*. Berlin: Julius Springer. (Translated as *Outlines of a Theory of the Light Sense*. Cambridge, MA: Harvard University Press, 1964.)

Holland, John H. 1998. *Emergence — From Chaos to Order*. Oxford: Oxford University Press.

Holt, P.J. 1976. "Causality and Our Conception of Matter." *Analysis* 37: 20–29.

Hooker, C.A. 1981. "Towards a General Theory of Reduction." *Dialogue* 20: 38–59, 201–36, 496–529.

Horgan, Terence. 1984. "Supervenience and Cosmic Hermeneutics." *Southern Journal of Philosophy*, supplementary volume, 22: 19–38.

———. 1989. "Mental Quausation." In *Philosophical Perspectives 3: Philosophy of Mind and Action Theory*, edited by J.E. Tomberlin, 47–76. Atascadero, CA: Ridgeview.

———. 1993. "From Supervenience to Superdupervenience: Meeting the Demands of a Material World." *Mind* 102: 555–86.

Hull, David. 1974. *The Philosophy of Biological Science*. Englewood Cliffs, NJ: Prentice Hall.

Hume, David. 1978. *A Treatise of Human Nature*. 2nd ed. Edited by L.A. Selby-Bigge and P.H. Nidditch. Oxford: Clarendon Press.

Humphreys, Paul W. 1989. "The Causes, Some of the Causes, and Nothing but the Causes." In *Minnesota Studies in the Philosophy of Science*, vol. 13: *Scientific Explanation*, edited by Philip Kitcher and Wesley C. Salmon, 283–306. Minneapolis: University of Minnesota Press.

———. 1996. "Aspects of Emergence." *Philosophical Topics* 24: 53–70.

———. 1997a. "How Properties Emerge." *Philosophy of Science* 64: 1–17.

———. 1997b. "Emergence, not Supervenience." *Philosophy of Science* 64 (proceedings): S337–45.

Huneman, Philippe. 2008. "Emergence Made Ontological? Computational versus Combinatorial Approaches." *Philosophy of Science* 75: 595–607.

Hurvich, L.M., and D. Jameson. 1957. "An Opponent-Process Theory of Color Vision." *Psychological Review* 64: 384–404.

Hüttemann, Andreas, and Orestis Terzidis. 2000. "Emergence in Physics." *International Studies in the Philosophy of Science* 14: 267–81.

Jackson, Frank. 1994. "Armchair Metaphysics." In *Philosophy in Mind: The Place of Philosophy in the Study of Mind*, edited by Michaelis Michael and John O'Leary-Hawthorne, 23–42. Dordrecht: Kluwer.

———. 1998. *From Metaphysics to Ethics*. Oxford: Clarendon Press.

Jacob, Pierre. 2002. "Some Problems for Reductive Physicalism." *Philosophy and Phenomenological Research* 65: 647–53.

James, William. 1890. *The Principles of Psychology*. 2 vols. New York: Henry Holt and Company. (Reprinted, New York: Dover, 1950.)

Johnson, Steven. 2001. *Emergence — The Connected Lives of Ants, Brains, Cities, and Software*. New York: Scribner.

Johnston, Mark. 1992. "How to Speak of the Colours." *Philosophical Studies* 68: 221–63.

Joseph, Geoffrey. 1980. “The Many Sciences and the One World.” *Journal of Philosophy* 77: 773–91.

Kamin, L.J. 1969. “Predictability, Surprise, Attention and Conditioning.” In *Punishment and Aversive Behavior*, edited by B.A. Campbell and R.M. Church, 279–96. New York: Appleton-Century-Crofts.

Kandel, Eric. 1995. “Cellular Mechanisms of Learning and Memory.” In *Essentials of Neural Science and Behavior*, edited by E.R. Kandel, J.H. Schwartz, and T.M. Jessell, 667–94. London: Prentice Hall.

———. 2000. “Cellular Mechanisms of Learning and the Biological Basis of Individuality.” In *Principles of Neural Science*, edited by E.R. Kandel, J.H. Schwartz, and T.M. Jessell, 1247–79. New York: McGraw-Hill.

Kaplan, David. 1989. “Demonstratives. An Essay on the Semantics, Logic, Metaphysics, and Epistemology of Demonstratives and Other Indexicals.” In *Themes from Kaplan*, edited by Joseph Almog, John Perry, and Howard Wettstein, 481–564. New York: Oxford University Press.

Keil, Geert. 2000. *Handeln und Verursachen*. Frankfurt am Main: Vittorio Klostermann.

Kemeny, John G., and Paul Oppenheim. 1956. “On Reduction.” *Philosophical Studies* 7: 6–19.

Kim, Jaegwon. 1973. “Causation, Nomic Subsumption and the Concept of Event.” *Journal of Philosophy* 70: 217–36. (Reprinted in Kim 1993b, 3–21.)

———. 1978. “Supervenience and Nomological Incommensurables.” *American Philosophical Quarterly* 15: 149–56.

———. 1984. “Concepts of Supervenience.” *Philosophy and Phenomenological Research* 45: 153–76. (Reprinted in Kim 1993b, 53–78.)

———. 1988a. “Explanatory Realism, Causal Realism, and Explanatory Exclusion.” *Midwest Studies in Philosophy* 12: 225–39.

———. 1988b. “Supervenience for Multiple Domains.” *Philosophical Topics* 16: 129–50. (Reprinted in Kim 1993b, 109–30.)

———. 1989a. “The Myth of Nonreductive Materialism.” *Proceedings and Addresses of the American Philosophical Association* 63: 31–47. (Reprinted in Kim 1993b, 265–84.)

———. 1989b. “Mechanism, Purpose and Explanatory Exclusion.” In *Philosophical Perspectives 3: Philosophy of Mind and Action Theory*, edited by James E. Tomberlin, 78–108. Atascadero, CA: Ridgeview. (Reprinted in Kim 1993b, 237–64.)

———. 1990. “Supervenience as a Philosophical Concept.” *Metaphilosophy* 21: 1–27. (Reprinted in Kim 1993b, 131–60.)

———. 1992a. “Multiple Realization and the Metaphysics of Reduction.” *Philosophy and Phenomenological Research* 52: 1–26. (Reprinted in Kim 1993b, 309–35.)

———. 1992b. “‘Downward Causation’ in Emergentism and Non-Reductive Physicalism.” In *Emergence or Reduction? Essays on the Prospects of Nonreductive Physicalism*, edited by Ansgar Beckermann, Hans Flohr, and Jaegwon Kim, 119–38. Berlin: de Gruyter.

———. 1993a. “Postscripts on Mental Causation.” In Kim 1993b, 358–67.

———. 1993b. *Supervenience and Mind*. Cambridge, UK: Cambridge University Press.

———. 1993c. "The Non-Reductivist's Troubles with Mental Causation." In *Mental Causation*, edited by John Heil and Alfred Mele, 189–210. Oxford: Clarendon Press. (Reprinted in Kim 1993b, 336–57.)

———. 1997a. "The Mind-Body Problem: Taking Stock after Forty Years." In *Philosophical Perspectives 11: Mind, Causation, and World*, edited by James E. Tomberlin, 185–207. Cambridge, MA: Blackwell.

———. 1997b. "Does the Problem of Mental Causation Generalize?" *Proceedings of the Aristotelian Society*, New Series, 97: 281–97.

———. 1998. *Mind in a Physical World*. Cambridge, MA: MIT Press.

———. 2002. "Responses." *Philosophy and Phenomenological Research* 65: 671–80.

———. 2005. *Physicalism, or Something Near Enough*. Princeton, NJ: Princeton University Press.

Kincaid, Harold. 1990. "Molecular Biology and the Unity of Science." *Philosophy of Science* 57: 575–93.

Kirk, Robert. 1996. "Strict Implication, Supervenience and Physicalism." *Australasian Journal of Philosophy* 74: 244–57.

———. 2001. "Non-Reductive Physicalism and Strict Implication." *Australasian Journal of Philosophy* 79: 544–52.

Kistler, Max. 1998. "Reducing Causality to Transmission." *Erkenntnis* 48: 1–24.

———. 1999a. "Causes as Events and Facts." *Dialectica* 53: 25–46.

———. 1999b. *Causalité et lois de la nature*. Paris: Vrin.

———. 1999c. "Multiple Realization, Reduction, and Mental Properties." *International Studies in the Philosophy of Science* 13: 135–49.

———. 2001. "Causation as Transference and Responsibility." In *Current Issues in Causation*, edited by Wolfgang Spohn, Marion Ledwig, and Michael Esfeld, 115–33. Paderborn: Mentis.

———. 2002a. "The Causal Criterion of Reality and the Necessity of Laws of Nature." *Metaphysica* 3: 57–86.

———. 2002b. "Erklärung und Kausalität." *Philosophia naturalis* 39, no. 1: 89–109.

———. 2002c. "Causation in Contemporary Analytical Philosophy." In *Causation*, vol. 2 of *Quaestio-Yearbook of the History of Metaphysics*, edited by C. Esposito and P. Porro, 635–68. Turnhout, Belgium: Brepols.

———. 2004a. "Causality in Contemporary Philosophy." *Intellectica* 38: 139–85.

———. 2004b. "Matérialisme et réduction de l'esprit." In *Les matérialismes (et leurs détracteurs)*, edited by Jean Dubessy, Guillaume Lecointre, and Marc Silberstein, 309–39. Paris: Syllepse. (Reprinted in *Matériaux philosophiques et pour un matérialisme contemporain: Sciences, ontologie, épistémologie*, edited by Marc Silberstein, 919–54. Paris: Éditions Matériologiques, 2013.)

———. 2005a. "Necessary Laws." In *Nature's Principles*, edited by Jan Faye, Paul Needham, Uwe Scheffler, and Max Urchs, 201–27. Dordrecht: Springer.

———. 2005b. "Lowe's Argument for Dualism from Mental Causation." *Philosophia: A Global Journal of Philosophy* 33: 319–29.

——. 2005c. "L'efficacité causale des propriétés dispositionnelles macroscopiques." In*Causes, Powers, Dispositions en philosophie: Le retour des vertus dormitives*, edited by Bruno Gnassounou and Max Kistler, 115–54. Paris: Presses Universitaires de France/Éditions ENS Rue d'Ulm.

——. 2005d. "Is Functional Reduction Logical Reduction?" *Croatian Journal of Philosophy* 5: 219–34.

——. 2006a. "La causalité comme transfert et dépendance nomique." *Philosophie* 89: 53–77.

——. 2006b. "Lois, exceptions et dispositions." In *Les dispositions en philosophie et en sciences*, edited by Bruno Gnassounou and Max Kistler, 175–94. Paris: CNRS Éditions. (English translation: "Laws, Exceptions, and Dispositions." *Journal for the Philosophy of Language, Mind and the Arts* 1, no. 1 [2020]: 45–66. https://edizionicafoscari.unive.it/media/pdf/journals/the-journal-for-the-philosophy-of-language-mind-an/2020/1/iss-1-1-2020_8T1ifxo.pdf.)

——. 2006c. "Les causes des actions." *Le temps philosophique, numéro spécial: L'action* 12: 141–75.

——. 2006d. *Causation and Laws of Nature*. London: Routledge.

——. 2007. "La réduction, l'émergence, l'unité de la science et les niveaux de réalité." *Matière première* 2: 67–97. (Reprinted in *Matériaux philosophiques et pour un matérialisme contemporain: Sciences, ontologie, épistémologie*, edited by Marc Silberstein, 179–212. Paris: Éditions Matériologiques, 2013.)

——. 2009. "Mechanisms and Downward Causation." *Philosophical Psychology* 22: 595–609.

——. 2011. "La causalité." In *Précis de philosophie des sciences*, edited by Anouk Barberousse, Denis Bonnay, and Mikaël Cozic, 100–40. Paris: Vuibert.

——. 2012. "Powerful Properties and the Causal Basis of Dispositions." In *Properties, Powers and Structures: Issues in the Metaphysics of Realism*, edited by Alexander Bird, Brian Ellis, and Howard Sankey, 119–37. New York: Routledge.

——. 2013. "The Interventionist Account of Causation and Non-Causal Association Laws." *Erkenntnis* 78: 65–84.

——. 2014. "Analysing Causation in Light of Intuitions, Causal Statements, and Science." In *Causation in Grammatical Structures*, edited by B. Copley and F. Martin, 76–99. Oxford: Oxford University Press.

——. 2016. "Espèces naturelles, profil causal et constitution multiple." *Lato sensu* 3, no. 1: 17–30. (Translation: "Natural Kinds, Causal Profile and Multiple Constitution." *Metaphysica* 19 [2018]: 113–35.)

——. 2017. "Higher-Level, Downward and Specific Causation." In *Philosophical and Scientific Pespectives on Downward Causation*, edited by Michele Paolini Paoletti and Francesco Orilia, 54–75. London: Routledge.

——. 2020. "Powers, Dispositions and Laws of Nature." In *Dispositionalism: Perspectives from Metaphysics and the Philosophy of Science*, edited by Anne Sophie Meincke, 171–88. Cham: Springer.

——. 2021. "Models of Downward Causation." In *Top-Down Causation and Emergence*, edited by Jan Voosholz and Markus Gabriel, 305–26. Cham: Springer.

———. 2022. "Lowe's Dualist Construal of Mental Causation." In *E.J. Lowe and Ontology*, edited by Mirosław Szatkowski, 239–59. New York: Routledge.

———. 2025. *Metaphysics of Causation*. Cambridge, UK: Cambridge University Press. DOI: https://doi.org/10.1017/9781009260800.

Kitcher, Philip. 1982. "Genes." *British Journal for the Philosophy of Science* 33: 337–59.

———. 1984. "1953 and All That: A Tale of Two Sciences." *Philosophical Review* 93: 335–73.

———. 1989. "Explanatory Unification and the Causal Structure of the World." In *Minnesota Studies in the Philosophy of Science 13: Scientific Explanation*, edited by Philip Kitcher and Wesley Salmon, 410–505. Minneapolis: University of Minnesota Press.

Klee, Robert. 1984. "Micro-Determinism and Concepts of Emergence." *Philosophy of Science* 51: 44–63.

Kripke, Saul A. 1972. "Naming and Necessity" and "Addenda to Saul A. Kripke's Paper 'Naming and Necessity." In *Semantics of Natural Language*, edited by D. Davidson and G. Harman, 253–355, 763–69. Dordrecht: Reidel. (Reprinted as a monograph: *Naming and Necessity*. Cambridge, MA: Harvard University Press, 1980.)

Kronz, Frederick M., and Justin T. Tiehen. 2002. "Emergence and Quantum Mechanics." *Philosophy of Science* 69: 324–47.

Krüger, Lorenz. 1989. "Reduction without Reductionism." In *An Intimate Relation*, edited by J.R. Brown and J. Mittelstrass, 369–90. Dordrecht: Kluwer.

Kuhn, Thomas. 1962. *The Structure of Scientific Revolutions*. Chicago: University of Chicago Press.

Kurtz, D.M. Jr. 1992. "Molecular Structure and Function Relationships of Hemerythrins." *Advances in Comparative and Environmental Physiology* 13: 151–71.

———. 1999. "Oxygen-Carrying Proteins: Three Solutions to a Common Problem." *Essays in Biochemistry* 34: 85–100.

Laplace, Pierre-Simon. 1825. *Essai philosophique sur les probabilités*. Reprinted, Paris: Christian Bourgeois, 1986.

Lennie, Peter. 2000. "Color Vision." In *Principles of Neural Science*, 4th ed., edited by E.R. Kandel, J.H. Schwartz, and T.M. Jessel, 572–89. New York: McGraw-Hill.

LePore, Ernest, and Barry Loewer. 1987. "Mind Matters." *Journal of Philosophy* 84: 630–42.

———. 1989. "More on Making Mind Matter." *Philosophical Topics* 17: 175–91.

Lewes, George H. 1875. *Problems of Life and Mind*. London: Trübner.

Lewis, David. 1973. *Counterfactuals*. Oxford: Basil Blackwell.

———. 1980. "Index, Content and Context." In *Philosophy and Grammar*, edited by Stig Kanger and Sven Öhman, 79–100. Dordrecht: Reidel.

———. 1983. "New Work for a Theory of Universals." *Australasian Journal of Philosophy* 70: 211–24.

———. 1986. *Philosophical Papers*. Vol. 2. New York: Oxford University Press.

———. 1994. "Reduction of Mind." In *A Companion to the Philosophy of Mind*, edited by Samuel Guttenplan, 412–31. Oxford: Blackwell.

———. 1997. "Finkish Dispositions." *Philosophical Quarterly* 47: 143–58.

Lipton, Peter. 1999. "All Else Being Equal." *Philosophy* 74: 155–68.

Lloyd Morgan, Conwy. 1926. *Emergent Evolution*. London: Williams and Norgate.

Locke, John. 1975. *An Essay Concerning Human Understanding*. Edited by Peter H. Nidditch. Oxford: Clarendon Press.

Loewer, Barry. 2002. "Comments on Jaegwon Kim's *Mind and the Physical World*." *Philosophy and Phenomenological Research* 65: 654–61.

Lowe, E.J. 1993. "The Causal Autonomy of the Mental." *Mind* 102: 629–44.

———. 1996. *Subjects of Experience*. Cambridge, UK: Cambridge University Press.

———. 2000a. "Causal Closure Principles and Emergentism." *Philosophy* 75: 571–85.

———. 2000b. *An Introduction to the Philosophy of Mind*. Cambridge, UK: Cambridge University Press.

———. 2001a. "Dispositions and Laws." *Metaphysica* 2: 5–23.

———. 2001b. "Event Causation and Agent Causation." *Grazer philosophische Studien* 61: 1–20.

Macdonald, Cynthia, and Graham Macdonald. 1986. "Mental Causes and Explanation of Action." In *Mind, Causation, and Action*, edited by Leslie Stevenson, Roger Squires, and John Haldane, 35–48. Oxford: Blackwell.

Machamer, Peter, Lindley Darden, and Carl F. Craver. 2000. "Thinking about Mechanisms." *Philosophy of Science* 67: 1–25.

Mackie, John L. 1965. "Causes and Conditions." *American Philosophical Quarterly* 2, no. 4: 245–55. (Reprinted in *Causation*, edited by Ernest Sosa and Michael Tooley, 33–55. Oxford: Oxford University Press, 1993.)

———. 1977. "Dispositions, Grounds, and Causes." *Synthese* 34: 361–70.

Malaterre, Christophe. 2010. *Les origines de la vie: Emergence ou explication réductive*. Paris: Hermann.

Malcolm, Norman. 1968. "The Conceivability of Mechanism." *Philosophical Review* 77: 45–72.

Malzkorn, Wolfgang. 2000. "Realism, Functionalism and the Conditional Analysis of Dispositions." *Philosophical Quarterly* 50: 452–69.

Marras, Ausonio. 1998. "Kim's Principle of Explanatory Exclusion." *Australasian Journal of Philosophy* 76: 439–51.

———. 2000. "Critical Note of Kim, *Mind in a Physical World*." *Canadian Journal of Philosophy* 30: 137–60.

———. 2002. "Kim on Reduction." *Erkenntnis* 57: 231–57.

Martin, Charles B. 1994. "Dispositions and Conditionals." *Philosophical Quarterly* 44: 1–8.

———. 1996. "Final Replies to Place and Armstrong." In *Dispositions: A Debate*, by David M. Armstrong, C.B. Martin, and U.T. Place, edited by Tim Crane, 163–92. London: Routledge.

Maull, Nancy. 1977. "Unifying Science without Reduction." *Studies in History and Philosophy of Science* 8: 143–62.

McCauley, Robert N. 1981. "Hypothetical Identities and Ontological Economizing: Comments on Causey's Program for the Unity of Science." *Philosophy of Science* 48: 218–27.

———. 1996. "Explanatory Pluralism and the Co-Evolution of Theories in Science." In *The Churchlands and Their Critics*, edited by Robert N. McCauley, 17–47. Cambridge, MA: Blackwell.

McLaughlin, Brian. 1989. "Type Epiphenomenalism, Type Dualism, and the Causal Priority of the Physical." In *Philosophical Perspectives 3: Philosophy of Mind and Action Theory*, edited by J.E. Tomberlin, 109–35. Atascadero, CA: Ridgeview.

———. 1992. "The Rise and Fall of British Emergentism." In *Emergence or Reduction? Essays on the Prospects of Nonreductive Physicalism*, edited by Ansgar Beckermann, Hans Flohr, and Jaegwon Kim, 49–93. Berlin: de Gruyter.

———. 1993. "On Davidson's Response to the Charge of Epiphenomenalism." In *Mental Causation*, edited by A. Mele and J. Heil, 27–40. Oxford: Oxford University Press.

———. 1995. "Varieties of Supervenience." In *Supervenience: New Essays*, edited by E. Savellos and Ü. Yalçin, 16–59. Cambridge, UK: Cambridge University Press.

Mellor, D.H. 1974. "In Defense of Dispositions." *Philosophical Review* 83: 157–81.

———. 2000. "The Semantics and Ontology of Dispositions." *Mind* 109: 757–80.

Menzies, Peter. 1988. "Against Causal Reductionism." *Mind* 97: 551–74.

———. 2008. "The Exclusion Problem, the Determination Relation, and Contrastive Causation." In *Being Reduced*, edited by Jakob Hohwy and Jesper Kallestrup, 196–217. Oxford: Oxford University Press.

Menzies, Peter, and Christian List. 2010. "The Causal Autonomy of the Special Sciences." In *Emergence in Mind*, edited by Cynthia Macdonald and Graham Macdonald, 108–28. Oxford: Oxford University Press.

Mill, John Stuart. 1843. *A System of Logic, Ratiocinative and Inductive*. London: Parker. (1891 edition reprinted, Honolulu: University Press of the Pacific, 2002. 1882 edition reprinted, https://www.gutenberg.org.)

Molnar, George. 1999. "Are Dispositions Reducible?" *Philosophical Quarterly* 49: 1–17.

———. 2003. *Dispositions: A Study in Metaphysics*. Oxford: Oxford University Press.

Morange, Michel. 1998. *La part des gènes*. Paris: Odile Jacob. (Translation: *The Misunderstood Gene*. Translated by Matthew Cobb. Cambridge, MA: Harvard University Press, 2001.)

Mossio, Matteo, and Jon Umerez. 2014. "Réductionnisme, holisme et émergentisme." In *Précis de philosophie de la biologie*, edited by T. Hoquet and F. Merlin, 179–89. Paris: Vuibert.

Mumford, Stephen. 1996. "Conditionals, Functional Essences and Martin on Dispositions." *Philosophical Quarterly* 46: 86–92.

———. 1998. *Dispositions*. Oxford: Oxford University Press.

Nagel, Ernest. 1951. "Mechanistic Explanation and Organismic Biology." *Philosophy and Phenomenological Research* 11: 327–38. (Reprinted in Nagel 1961, 398–446.)

———. 1952. "Wholes, Sums and Organic Unities." *Philosophical Studies* 3: 17–32. (Reprinted in Nagel 1961, 380–97; reprinted in *Parts and Wholes*, edited by Daniel Lerner, 135–55. New York: Free Press, 1963.)

———. 1961. *The Structure of Science*. London: Routledge and Kegan Paul.

Nagel, Thomas. 1979. "Panpsychism." In *Mortal Questions*, edited by T. Nagel, 181–95. Cambridge, UK: Cambridge University Press.

Needham, Paul. 2000. "What Is Water?" *Analysis* 60: 13–21.

Newman, David V. 1996. "Emergence and Strange Attractors." *Philosophy of Science* 63: 245–61.

———. 2001. "Chaos, Emergence, and the Mind-Body Problem." *Australasian Journal of Philosophy* 79: 180–96.

Nickles, Thomas. 1973. "Two Concepts of Intertheoretic Reduction." *Journal of Philosophy* 70: 181–201.

Noordhof, Paul. 1998. "Do Tropes Resolve the Problem of Mental Causation?" *Philosophical Quarterly* 48: 221–26.

O'Connor, Timothy. 1994. "Emergent Properties." *American Philosophical Quarterly* 31: 91–104.

O'Regan, Kevin J., and Alva Noë. 2001. "What It Is Like to See: A Sensorimotor Theory of Perceptual Experience." *Synthese* 129: 79–103.

Pap, Arthur. 1951–52. "The Concept of Absolute Emergence." *British Journal for the Philosophy of Science* 2: 302–11.

Papineau, David. 1993. *Philosophical Naturalism*. Oxford: Blackwell.

Pereboom, D., and Kornblith, H. 1991. "The Metaphysics of Irreducibility." *Philosophical Studies* 63: 125–45.

Pietroski, Paul. 1994. "Mental Causation for Dualists." *Mind and Language* 9: 336–66.

Pietroski, Paul, and Georges Rey. 1995. "When Other Things Aren't Equal: Saving *Ceteris Paribus* Laws from Vacuity." *British Journal for the Philosophy of Science* 46: 81–110.

Popper, Karl R. 1957. "The Aim of Science." Reprinted in Karl R. Popper, *Objective Knowledge*, 191–205. Oxford: Oxford University Press. (Earlier version published in *Ratio* 1: 24–35).

———. 1977. "Natural Selection and the Emergence of Mind." *Dialectica* 32: 339–55.

Price, H.H. 1953. *Thinking and Experience*. London: Hutchinson.

Prior, Elizabeth. 1985. *Dispositions*. Aberdeen: Aberdeen University Press.

Prior, Elizabeth W., Robert Pargetter, and Frank Jackson. 1982. "Three Theses about Dispositions." *American Philosophical Quarterly* 19: 251–57.

Putnam, Hilary. 1963. "Brains and Behavior." In *Analytical Philosophy, Second Series*, edited by Ronald J. Butler, 211–35. Oxford: Blackwell. (Reprinted in Hilary Putnam, *Mind, Language, and Reality: Philosophical Papers*, vol. 2, 325–41. Cambridge, UK: Cambridge University Press, 1975.)

———. 1965. "How not to Talk about Meaning: Comments on J.J.C. Smart." In *Boston Studies in the Philosophy of Science*, vol. 2, *In Honor of Philipp Frank*, edited by Robert S. Cohen and Marx R. Wartofsky, 205–22. New York: Humanities Press. (Reprinted in Hilary Putnam, *Mind, Language, and Reality: Philosophical Papers*, vol. 2, 117–31. Cambridge, UK: Cambridge University Press, 1975.)

———. 1967. "The Nature of Mental States." First published as "Psychological Predicates." In *Art, Mind, and Religion*, edited by William H. Capitan and Daniel D. Merrill, 37–48. Pittsburgh: University of Pittsburgh Press. (Reprinted in Hilary Putnam,

Mind, Language, and Reality: Philosophical Papers, vol. 2, 429–40. Cambridge, UK: Cambridge University Press, 1975.)

———. 1975a. "The Meaning of 'Meaning.'" In *Language, Mind and Knowledge: Minnesota Studies in the Philosophy of Science*, vol. 7, edited by Keith Gunderson, 131–93. Minneapolis: University of Minnesota Press. (Reprinted in Hilary Putnam, *Mind, Language and Reality: Philosophical Papers*, vol. 2, 215–71. Cambridge, UK: Cambridge University Press, 1975.)

———. 1975b. "Philosophy and Our Mental Life." In *Mind, Language and Reality: Philosophical Papers*, vol. 2, by Hilary Putnam, 291–303. Cambridge, UK: Cambridge University Press.

———. 1992. *Renewing Philosophy*. Cambridge, MA: Harvard University Press.

Quine, Willard V.O. 1939. "Designation and Existence." *Journal of Philosophy* 36: 701–09.

———. 1948. "On What There Is." *Review of Metaphysics* 2, no. 1: 21–38. (Reprinted in Willard V.O. Quine, *From a Logical Point of View*, 1–19. New York: Harper, 1953.)

———. 1971. *The Roots of Reference*. LaSalle, IL: Open Court.

———. 1976. "A Logistical Approach to the Ontological Problem." In Willard V.O. Quine, *The Ways of Paradox and Other Essays*, rev. and enlarged ed., 197–202. Cambridge, MA: Harvard University Press.

Reif, Frederick. 1967. *Statistical Physics*. Berkeley Physics Course, vol. 5. New York: McGraw-Hill.

Rescorla, Robert A. 1968. "Probability of Shock in the Presence and Absence of CS in Fear Conditioning." *Journal of Comparative and Physiological Psychology* 66: 1–5.

———. 1988. "Pavlovian Conditioning: It's not What You Think It Is." *American Psychologist 43*: 151–60.

Rescorla, Robert A., and A.R. Wagner. 1972. "A Theory of Pavlovian Conditioning: Variations in the Efficaciousness of Reinforcement and Nonreinforcement." In *Classical Conditioning 2: Current Research and Theory*, edited by A.H. Black and W.F. Prokasy, 64–99. New York: Appleton-Croft-Century.

Rey, Georges. 1997. *Contemporary Philosophy of Mind*. Oxford: Blackwell.

Richardson, Robert. 1979. "Functionalism and Reductionism." *Philosophy of Science* 46: 533–58.

Richet, Pascal. 2001. *The Physical Basis of Thermodynamics, with Applications to Chemistry*. New York: Springer.

Robb, David. 1997. "The Properties of Mental Causation." *Philosophical Quarterly* 47: 178–94.

Robinson, Howard. 1982. *Matter and Sense*. Cambridge, UK: Cambridge University Press.

Rosenberg, Alexander. 1985. *The Structure of Biological Science*. Cambridge, UK: Cambridge University Press.

Rueger, Alexander. 2000a. "Robust Supervenience and Emergence." *Philosophy of Science* 67: 466–89.

———. 2000b. "Physical Emergence, Diachronic and Synchronic." *Synthese* 124, no. 3: 297–322.

———. 2001. "Explanations at Multiple Levels." *Minds and Machines* 11: 503–20.

———. 2004. "Reduction, Autonomy and Causal Exclusion among Physical Properties." *Synthese* 139, no. 1: 1–21.

Ryle, Gilbert. 1949. *The Concept of Mind*. London: Hutchinson. (Reprinted, sixtieth anniversary ed. Abingdon: Routledge, 2009.)

Salmon, Wesley. 1990. *Four Decades of Scientific Explanation*. Minneapolis: University of Minnesota Press.

Schaffer, Jonathan. 2014. "The Metaphysics of Causation." In *Stanford Encyclopedia of Philosophy*. http://plato.stanford.edu/entries/causation-metaphysics/.

Schaffner, Kenneth. 1967. "Approaches to Reduction." *Philosophy of Science* 34: 137–47.

———. 1976. "Reductionism in Biology: Prospects and Problems." In *PSA: Proceedings of the Biennial Meeting of the Philosophy of Science Association 1974*, edited by R.S. Cohen et al., 613–32. Dordrecht: Reidel.

———. 1992. "Philosophy of Medicine." In *Introduction to the Philosophy of Science*, edited by Merrilee Salmon et al., 310–45. Indianapolis: Hackett.

———. 1993. *Discovery and Explanation in Biology and Medicine*. Chicago: University of Chicago Press.

Schiffer, Stephen. 1987. *Remnants of Meaning*. Cambridge, MA: MIT Press.

Schrenk, Markus. 2006. *Les dispositions en philosophie et en sciences*. In *Les capacités peuvent-elles nous sauver des lois* ceteris paribus, edited by Bruno Gnassounou and Max Kistler, 147–73. Paris: CNRS Éditions.

Schröder, Jürgen. 1998. "Emergence: Non-Deducibility or Downwards Causation?" *Philosophical Quarterly* 48: 433–52.

Schurz, Gerhard. 2002. "*Ceteris paribus* Laws: Classification and Destruction." *Erkenntnis* 57: 351–72.

Searle, John. 1992. *The Rediscovery of the Mind*. Cambridge, MA: MIT Press.

Shepard, Roger. 1962. "The Analysis of Proximities: Multidimensional Scaling with an Unknown Distance Function." *Psychometrika* 27: 125–40, 219–46.

———. 1965. "Approximation to Uniform Gradients of Generalization by Monotone Transformations of Scale." In *Stimulus Generalization*, edited by David I. Mostofsky, 95–110. Stanford, CA: Stanford University Press.

———. 1974. "Representation of Structure in Similarity Data: Problems and Prospects." *Psychometrika* 39: 373–421.

Shoemaker, Sydney. 1980. "Causality and Properties." In *Time and Cause: Essays Presented to Richard Taylor*, edited by Peter van Inwagen, 109–35. Dordrecht: Reidel. (Reprinted in Sydney Shoemaker, *Identity, Cause and Mind*, 206–33. Cambridge, UK: Cambridge University Press, 1984; reprinted in *Properties*, edited by D.H. Mellor and A. Oliver, 228–54. Oxford: Oxford University Press, 1997.)

———. 1998. "Causal and Metaphysical Necessity." *Pacific Philosophical Quarterly* 79: 59–77.

Sklar, Lawrence. 1993. *Physics and Chance*. Cambridge, UK: Cambridge University Press.

Slors, Marc. 1998. "Two Claims that Can Save a Nonreductive Account of Mental Causation." In *Human Action, Deliberation, and Causation*, edited by Jan Bransen and Stefaan E. Cuypers, 225–48. Dordrecht: Kluwer.

Sosa, Ernest. 1984. "Mind-Body Interaction and Supervenient Causation." *Midwest Studies in Philosophy* 9: 271–81.

Sperry, Roger W. 1964. "The Great Cerebral Commissure." *Scientific American* 210: 42–52.

———. 1969. "A Modified Concept of Consciousness." *Psychological Review* 76: 532–36.

———. 1976. "Mental Phenomena as Causal Determinants in Brain Function." In *Consciousness and the Brain*, edited by Gordon G. Globus, Grover Maxwell, and Irwin Savodnik, 163–77. New York: Plenum Press.

———. 1986. "Macro- versus Micro-Determinism." *Philosophy of Science* 53: 265–70.

———. 1992. "Turnabout on Consciousness: A Mentalist View." *Journal of Mind and Behavior* 13: 259–80.

Spohn, Wolfgang. 1997. "Begründungen a priori — oder: Ein frischer Blick auf Dispositionsprädikate." In *Das weite Spektrum der analytischen Philosophie: Festschrift für Franz von Kutschera*, edited by Wolfgang Lenzen, 323–45. Berlin: de Gruyter.

Squire, Larry R., and Eric R. Kandel. 1999. *Memory: From Mind to Molecules*. New York: Scientific American Library.

Squires, Roger. 1968. "Are Dispositions Causes?" *Analysis* 29: 45–47.

Stalnaker, Robert. 1978. "Assertion." In *Syntax and Semantics*, vol. 9, edited by Peter Cole, 315–32. New York: Academic Press.

———. 1993. "Twin Earth Revisited." *Proceedings of the Aristotelian Society* 93: 297–311.

Stegmüller, Wolfgang. 1983. *Probleme und Resultate der Wissenschaftstheorie und Analytischen Philosophie, Vol. I: Erklärung, Begündung, Kausalität*. 2nd ed. Berlin: Springer.

Stephan, Achim. 1992. "Emergence — A Systematic View on Its Historical Facets." In *Emergence or Reduction? Essays on the Prospects of Nonreductive Physicalism*, edited by Ansgar Beckermann, Hans Flohr, and Jaegwon Kim, 25–48. Berlin: de Gruyter.

———. 1999. *Emergenz: Von der Unvorhersagbarkeit zur Selbstorganisation*. Dresden: Dresden University Press.

———. 2006. "The Dual Role of 'Emergence' in the Philosophy of Mind and in Cognitive Science." *Synthese* 151, no. 3: 485–98.

Steward, Helen. 1997. *The Ontology of Mind*. Oxford: Oxford University Press.

Strawson, Peter F. 1959. *Individuals*. London: Methuen.

Suppes, Patrick. 1967. "What Is a Scientific Theory?" In *Philosophy of Science Today*, edited by Sidney Morgenbesser, 55–67. New York: Basic Books.

Teller, Paul. 1992. "A Contemporary Look at Emergence." In *Emergence or Reduction? Essays on the Prospects of Nonreductive Physicalism*, edited by Ansgar Beckermann, Hans Flohr, and Jaegwon Kim, 139–53. Berlin: de Gruyter.

Thomasson, Amie. 1998. "A Nonreductivist Solution to Mental Causation." *Philosophical Studies* 89: 181–95.

Thompson, Ian J. 1988. "Real Dispositions in the Physical World." *British Journal for the Philosophy of Science* 39: 67–79.

Tobin, Emma. 2010. "Microstructuralism and Macromolecules: The Case of Moonlighting Proteins." *Foundations of Chemistry* 12: 41–54.

Tompa, P., C. Szasz, and L. Buday. 2005. "Structural Disorder Throws New Light on Moonlighting." *Trends in Biochemical Sciences* 30: 484–89.

Tully, Robert E. 1981. "Emergence Revisited." In *Pragmatism and Purpose: Essays Presented to T.A. Goudge*, edited by L.W. Summer, J.G. Slater, and F. Wilson, 261–77. Toronto: University of Toronto Press.

van Cleve, James. 1990. "Mind — Dust or Magic? Panpsychism versus Emergence." In *Action Theory and Philosophy of Mind*, vol. 4 of *Philosophical Perspectives*, edited by J. Tomberlin, 215–26. Atascadero, CA: Ridgeview.

van Eck, Dingmar, Huib Looren de Jong, and Maurice K.D. Schouten. 2006. "Evaluating New Wave Reductionism: The Case of Vision." *British Journal for the Philosophy of Science* 57: 167–96.

van Holde, K.E., and Karen I. Miller. 1995. "Hemocyanins." *Advances in Protein Chemistry* 47: 1–81.

van Holde, K.E., K.I. Miller, and H. Decker. 2001. "Hemocyanins and Invertebrate Evolution." *Journal of Biological Chemistry* 276, no. 19: 15563–66.

Wimsatt, William C. 1976a. "Reductionism, Levels of Organization, and the Mind-Body Problem." In *Consciousness and the Brain*, edited by Gordon G. Globus, Grover Maxwell, and Irwin Savodnik, 205–67. New York: Plenum Press.

———. 1976b. "Reductive Explanation: A Functional Account." In *PSA: Proceedings of the Biennial Meeting of the Philosophy of Science Association 1974*, edited by R.S. Cohen et al., 671–710. Dordrecht: Reidel.

———. 1986. "Forms of Aggregativity." In *Human Nature and Natural Knowledge*, edited by A. Donagan, A.N. Perovich Jr., and M.V. Wedin, 259–91. Boston Studies in the Philosophy of Science 89. Dordrecht: Reidel.

———. 1996. "Aggregativity: Reductive Heuristics for Finding Emergence." *Philosophy of Science* 64 (proceedings): S372–84.

Witmer, D. Gene. 2003. "Functionalism and Causal Exclusion." *Pacific Philosophical Quarterly* 84: 198–214.

Wittgenstein, Ludwig. 1958. *Philosophical Investigations*. 2nd ed. Oxford: Blackwell.

Wong, Hong Yu. 2006. "Emergents from Fusion." *Philosophy of Science* 73: 345–67.

Woodger, J.H. 1952. *Biology and Language*. Cambridge, UK: Cambridge University Press.

Woodward, James. 2003. *Making Things Happen*. Oxford: Oxford University Press.

———. 2010. "Causation in Biology: Stability, Specificity, and the Choice of Levels of Explanation." *Biology and Philosophy* 25: 287–318.

Worley, Sara. 1997. "Determination and Mental Causation." *Erkenntnis* 46: 281–304.

Yablo, Stephen. 1992. "Mental Causation." *Philosophical Review* 101: 245–80.

———. 1997. "Wide Causation." In *Philosophical Perspectives 11: Mind, Cause and World*, edited by James E. Tomberlin, 251–81. Cambridge, MA: Blackwell.

Zeki, Semir. 1993. *A Vision of the Brain*. London: Blackwell.

Index

Kolibri 360 GmbH
Ferdinand-Jühlke-Straße 7
99095 Erfurt, Deutschland
produktsicherheit@kolibri360.de